Interacting with Animals

Understanding their Behaviour and Welfare

W0235327

FSC
www.fsc.org

MIX

Paper from
responsible sources

FSC® C022174

This book is a revised translation of the French publication *Vivre parmi les animaux, mieux les comprendre*, authored by Pierre Le Neindre and Bertrand L. Deputte.

Interacting with Animals

Understanding their Behaviour and Welfare

Pierre Le Neindre

Institut national de la recherche agronomique (INRA)
France

and

Bertrand Deputte

École nationale vétérinaire d'Alfort (ENVA)
France

Translated by David Lindsay

The University of Western Australia (UWA)
Australia

CABI is a trading name of CAB International

CABI
Nosworthy Way
Wallingford
Oxfordshire OX10 8DE
UK

CABI
200 Portland Street
Boston
MA 02114
USA

Tel: +44 (0)1491 832111
E-mail: info@cabi.org
Website: www.cabi.org

T: +1 (617)682-9015
E-mail: cabi-nao@cabi.org

Originally published in French under the title *Vivre parmi les animaux, mieux les comprendre* by Pierre Le Neindre and Bertrand L. Deputte.
© Éditions Quæ, 2020

A catalogue record for this book is available from the British Library, London, UK.

ISBN-13: 9781800622388 (Hardback)
          9781800622395 (Paperback)
          9781800622401 (eBook)
          9781800622418 (ePub)

Commissioning Editor: Caroline Makepeace
Editorial Assistant: Lauren Davies
Production Editor: James Bishop
Translators: David Lindsay

Typeset by Exeter Premedia Services Pvt Ltd, Chennai, India
Printed and bound in the UK by Severn, Gloucester

Contents

Contributors

Bertrand Deputte, École nationale vétérinaire d'Alfort (ENVA), 7 avenue du Général de Gaulle, 94700 Maisons-Alfort, France. Email: bert.deputte@wanadoo.fr

Pierre Le Neindre, Institut national de la recherche agronomique (INRA), 147 rue de l'Université, 75338 Paris, Cedex 07, France. Email: Pierre.le-Neindre@orange.fr

David Lindsay, The University of Western Australia (UWA), 35 Stirling Highway, Perth, WA 6009, Australia. Email: david.lindsay@uwa.edu.au and dotncarry7@gmail.com

Preface

Scientific knowledge is a cumulative 'adventure'. Research at a given time, even if it is in response to a fad, is based, and should be based on knowledge that has come before it. However, the need imposed on researchers to 'publish or perish' means that the use of previous work tends to focus on a profusion of 'small' more recent articles published within the last decade. We, who have worked through several decades, are seeing that many original works are no longer being cited. So, instead of seeing an evolution of concepts based on accumulated, verifiable information, we see a weakening or, even worse, a deterioration of concepts and a profusion of 'facts' that are difficult to put into sensible context. This profusion of facts about many species gives the illusion we know a lot about them but does little to improve our understanding of how they live. It requires time – a lot of time – and reflection to make worthwhile analyses. In our favour, we now have the advantage that most texts are digitized and can be widely accessed. One can easily find earlier articles, dating back several centuries. When one does this, it becomes clear that many more recent reviews, considered triumphantly as innovative, are little more than a rework of much more advanced thoughts of former scientists. We have rehabilitated these thoughts from time to time in this book, to emphasize the 'rediscovery of the wheel'. We have done this to show that behaviour in non-human animals is not simply a matter of anecdotes but, more beneficially, of scientific research based on methodologies adapted to questions being asked.

The accumulation of new data is not that interesting unless we back our conclusion on previous data. As was said, among others by Newton in 1686 we are dwarf but we can see further when sitting on the shoulders of giants.

Pierre Le Neindre,
Bertrand Deputte and
David Lindsay
May 2022

Introduction

Vertebrates of all types, whether they have hair, feather, scales or whatever are all around us and part of our everyday life. But do we really know them for what they are or just for what we expect them to be? This book presents an insight into animal behaviour in a way that shows the reader what animals do or are capable of doing to enable them to thrive in complex environments and, in particular, in the anthropogenic environment they share with humans. This is the starting point from which we can begin to define how we wish to live with animals in the future.

We are constantly conscious of animals in our day-to-day lives, either directly or indirectly. Some live with us in our houses. Some we eat. Others we hunt for food or cull them if we consider them to be pests. Still others fill additional important roles in our lives. For example, we ride horses, we have guide dogs and dogs that detect dangerous substances – or even pleasant ones like truffles. They are present in our books and our movies. They are symbolized in toys. Who has not been worried after finding that the teddy bear of our childhood was much more than a mere toy but an image of a beast living in the wild and in danger of extinction?

Many people want to give authoritative advice on how we should handle and treat our animals. Some think we shouldn't even be asking questions about the place of animals in our environment. They criticize our being sentimental about animals when many humans are starving. Others champion 'animal liberation' claiming that they should be left to have a free and full life. But to which species is this 'liberation' addressed? At the extremes, some people consider animals to be simply a source of food, or competitors, or parasites that put our food resources or health in danger, while others believe that the value of the lives of animals is every bit as important as that of humans. These issues are perturbing because they lead us to consider seriously where humans fit within the global biome. Some people may think that is not important but, surely, we need to question our relationships with non-humans on a rational and scientific basis.

Despite their constant presence in our thinking and our literature and our discussions, do we really understand the animals? Who are they? Have we really spent enough time observing them from a point of view that doesn't first question how they benefit us? Similarly, we sometimes judge the caretakers of animal welfare without knowing their real motivations.

During our careers in research, Pierre Le Neindre studied the behaviour of sheep and cattle, Bertrand Deputte specialized in primates, dogs and cats, and David Lindsay studied the behaviour of sheep. Despite all the time spent with animals, we still question if we really know them, if we understand what they expect and how they perceive what is important for them. Apart from understanding animals as metabolic 'machines', as some have described them, have we been, or should we be, interested in their mental states? Are words such as 'desire', 'suffering' and 'attachment', used commonly when characterizing human behaviour, meaningful when considering animals?

We are aware that our thinking is rooted in a West European context and background. However, how would we tackle the question if we had lived in different environments? If we were, for example, Fulani pastoralists, Australian hunter gatherers or Inuit? Closer to home, how would we converse with somebody who never interacted with a cow except when they see it as a piece of meat under cling wrap at the supermarket or from a movie about the Peul tribe roaming on the Serengeti plains? How can we understand those who adore their dogs and cats at the same time as they are unaware of the homeless sheltering under the porch of their front door? Can we even understand the differences between a breeder of Salers, a hardy breed of cattle in the Auvergne mountains of France, and a breeder of Holsteins on a farm in Brittany fully equipped with modern technologies such as feed dispensers, milking robots and automatic detectors of lameness and oestrus? Of course, much of these involve societal issues but we have confined ourselves in this book to questions of biology, so we use ethological scientific evidence about behaviour of animals in their habitual environments as farm or pet animals. Some of that evidence comes from our own experience but most comes from the vast scientific literature that takes advantage of the collective expertise from other stimulating and productive fields such as psychophysiology, neurophysiology and neurobiology. But we have largely limited the material in this book to analysing ethological results in which we believe we have expertise.

The basic question here is, 'Do we understand animals?' To try to answer this we must first of all define what we call an animal. For the systematists the animal world is very large, but it is differentiated from the plants only by their cellular constitution and their chemical composition. The *minimum* definition is that they are living beings devoid of both plastids and cellulose. Even mobility and ability to feel cannot be used for differentiating animals from the other living creatures (Bertin, 1949), so chimpanzee and sponges fit under the all-embracing word, 'animal'. However, within this large and diverse spectrum we will concentrate mainly on vertebrates because we know them better and not because the same issues are not relevant to non-vertebrates.

Reference is frequently made to animals being inferior or superior. This classification is problematic because, very often, the inferior or superior character is defined by taking Western adult humans as a reference. Even the differences between animals that seem familiar are difficult to characterize. So, for example, what distinguishes a lion and a lamb, a Holstein dairy cow, living in modern dairy farms and a Camargue cow living in the marshes of the Rhône delta, or between a fighting dog and a Chihuahua? Even the word 'animal' poses problems. Descartes (1596–1650) described the human as an animal. For him, a human was an animal, and what we would normally describe as an animal, he called a 'beast' to differentiate it semantically from a human (Descartes, 1637). A special animal certainly ... but still an animal that he semantically distinguishes from a beast!

The next important question is: 'Do we have robust objective tools for describing the behaviour of animals?' Do we have objective epistemological concepts and words for characterizing animals and their behaviour? Do we have an adequate scientific vocabulary that describes pertinent and not so pertinent behaviour? To understand the behaviour of animals we have to observe them with a naturalist approach in a range of natural and experimental contexts. Experience also shows that the experimental set-up must be adapted to the species to obtain answers that are trustworthy and useful.

Humans must also have the humility to consider that the answers that animals can give, and that humans may understand, are never fixed and always open to questions even if they are intellectually stimulating. This relative and provisional process is at the essence of knowledge in biology of behaviour. In our mind it is the only way to generate interpretable and irrefutable answers. Just because humans have the language to do so, doesn't mean that they are legitimate spokespersons to speak on the name of the animals or that they can state what they are and what they want.

Imagine a world map. On that map, where you stand is marked with a sharp pin saying, 'you are here on this map'. That image represents how much *Homo sapiens* count among the whole animal kingdom – in fact very little! As humans we are the result of evolution and natural selection. As brilliantly stated by Tattersall (2002, p. 91) 'species come, species go and, once in the long while something unexpected happens', so our curiosity and wonder should not be limited by specific frontiers. Whatever the species, ways to adapt are innumerable and often extraordinary. We certainly do not share the view of P. Teilhard de Chardin (1956) who wrote that the ultimate creation of human was through evolution and for God's purpose. Theology is not under our rationality.

Our curiosity and the open-mindedness that must accompany it must remain intact to ensure that we respect all species. As Budiansky (1998,

p. 194) rightly points out: 'All [creatures] hit upon unique ways to make a living against all probabilities. And that is something to respect and to treasure.'

Our interest and wonder must know no specific boundaries, the adaptations are immensely varied and often extraordinary, whatever the nature of the species: 'It is folly, and of an anthropomorphism of the worst kind, to insist that, to be truly wonderful, it [intelligence] must the same as ours' (Budiansky, 1998, p. 194).

H. sapiens was one of those 'unexpected things' that emerged in the same way as the elephant, the naked mole rat, the hammerhead shark, the southern kiwi, the emperor penguin, the earthworm, the slave ants, etc. It is this biological reality that we wish to elicit, so that we can deal with the anthropocentric and generally triumphalist discourses that are legion, with their qualifiers of 'augmented man' and 'artificial intelligence' which are in fact technological products of our intelligence and our fantasies.

In the course of evolution, millions of animal species have appeared, millions have disappeared, millions remain – all this through natural selection. *H. sapiens* is one of the products of this natural selection. But this species alone was able to 'adapt' other species particularly since the 19th century and the time of Lamarck and Darwin, to create, if not species, at least 'phenotypes' for its own benefits through artificial selection. But the species that we call 'domestic' are very few in number compared with all the species that appeared by natural selection. But, for *H. sapiens*, it was a big step.

As biologists, our view of the animal kingdom is largely panoramic rather than 'navel-gazing'. We marvel at the living. For example, birdwatchers are amazed at the richness of the different bird species. But, in reality, only the behaviour of few species has been superficially scientifically studied. We strive to produce scientific knowledge about the behaviour of some animal species, especially how they live. Even when they have been described we must be careful to present our findings on a rational basis. To fulfil that objective, we should never take for granted what we, as humans, think what the animals are thinking but we should find ways to question them so that we can understand them better. That should be our strategy because if it isn't, we will be constantly falling into traps when we make decisions about how we treat animals with which we have contact.

One of those traps is to treat animals as mere mechanics with only an instrumental value. In that case the care given to them is only related to the benefit we expect from them. Another trap is to consider it will never be possible to evaluate their desires and expectations, which is a very paradoxical scientific concept and a lack of faith in the imagination of the future scientists. Another is to consider that animals do not suffer or have desires. If that is the case, understanding them will be almost impossible. As a result, some people believe that the only solution is not to interfere with them in their 'natural environment'.

We consider that we are involved directly or indirectly in a network of relationships with animals. Our actions can have positive and negative consequences for them or their environments where we also live. We must take care of our 'room-mates' because we attach an ethical value to the existence of these animals.

One goal, which is becoming increasingly difficult to reach due to development of new technologies is to design a living environment that is sustainable and allows them a good life. Understanding them better is absolutely essential to that goal.

Our book therefore reviews the diverse aspects of animal behaviour, concentrating largely on those animals for which humans have a responsibility – so-called domestic animals. We emphasize the consequences of these multiple elements on how we interact with these animals in terms of both well-being and relations with humans and in particular with humans' role as caretakers.

References

Bertin, L. (1949) *La Vie des Animaux*, tome 1. Librairie Larousse, Paris, 493 pp.

Budiansky, S. (1998) *If a Lion Could Talk: How Animals Think*. Weidenfeld and Nicolson, London, 219 pp.

Descartes, R. (1637) [1982] *Discours de la Méthode*, 1982 10/18 edn. Librairie philosophique J. Vrin, Paris.

Tattersall, I. (2002) *The Monkey in the Mirror*. Oxford University Press, Oxford, 220 pp.

Teilhard de Chardin, P. (1956) *Le Phénomène Humain*. Les Éditions du Seuil, Paris, 348 pp.

Ethology, the Science of Understanding Animals

1.1 An Ancient and Largely Unknown Science

Our knowledge of animal species stems from the curiosity of naturalists dating back as far as Aristotle. Aristotle, in his *Histoire des animaux* ('History of animals') (translated by Louis, 1969) described not only the characteristics of the anatomy of many animal species, but also their behaviour. His observations are not detailed, but they are precise enough to make it clear that Aristotle was a keen observer of the species he described. Much later, Descartes (1596–1650; see Descartes, 1637 [1982 synthesis]) and many of his followers saw animals only as 'automata' that followed a predetermined programme. At a philosophical level, Descartes suggested that there was a large gap between the human species and 'animals', based on his belief that animals didn't have a soul like humans. He also believed that both humans and animals functioned similarly to automatons and this view prevailed for several centuries and is still the subject of hot debate in some quarters.

However, around the same time as Descartes, in 1647, Cureau de la Chambre, 1647 (1596–1669) published an explicit work on how people in that era thought about animals, under the title, *Traité de la connoissance des animaux, où tout ce qui a esté dict pour et contre le raisonnement des bestes est examine* ('Treatise on the knowledge of animals, where everything that has been said for and against the reasoning of the beasts is examined'). Cureau de la Chambre challenged Chanet (1646) who defended Descartes' 'mechanistic' approach. He argued that what some people considered to be instinctive, automatic, stereotyped behaviour in higher vertebrates like mammals and birds, involved processes that are now referred to as cognition or thought. Further, he argued that cognition involved memory, and a range of different ways of learning. For example, a dog that has seized prey gets satisfaction from ingesting it, so that when it sights the same sort of prey again it will begin to hunt it. Cureau de la Chambre also described exactly, in 17th-century terms, what Marler and Hamilton (1966) described in the 20th century as rhythmic behaviours. In other words, behaviour such

© CAB International 2022. *Interacting with Animals: Understanding their Behaviour and Welfare* (Pierre Le Neindre and Bertrand L. Deputte) DOI: 10.1079/9781800622418.0001

as hunger is induced by internal stimuli that lead to a phase of 'appetite', then a phase of consumption (ingestion of the prey) and a phase of satiety. But Cureau de la Chambre, a physician and philosopher, wrote more as a learned thinker who based his thoughts on the more or less complex behaviours of known animals, than as a naturalist questioning the apparent simplicity of the behaviour of the species he studied.

A century after Cureau de la Chambre, Leroy, in his *Lettres sur les animaux*, described the behaviour of certain animal species similarly (Leroy, 1768). He drew his observations from his role as manager of royal hunting parties. He observed only a few species of mammals and birds, but Leroy introduced for the first time the notion of adaptation. He criticized Descartes' view that animals were behavioural automatons: 'Explanations of the common things that animals do every day are quite contrary to the views of supporters of automatism' (Leroy, 1768, p. 5 and p. 148). The complexity of the behaviour that he saw even in the everyday life of animals led him to refute any concept of instinct that implied stereotyped behaviour by all individuals of the same species. He envisaged that behaviour is controlled by a form of intelligence based on learning and involving memory. This learning, according to Leroy, allows a flexibility of behaviour that is not possible with an automatism or with instinct.

Leroy focused more on carnivores, and therefore predators, than on other species. He reasoned that, because prey behave in many ways, carnivores had to develop anti-predator strategies to combat them. As an extension to this way of thinking, he asserted that the 'experience of carnivores' must be much superior to that of frugivores, or fruit-eating species. So, Leroy raised the idea of a difference between a non-social cognition, applying to inanimate objects, and a social cognition or a non-social cognition that considers mobile, and essentially variable and unpredictable 'objects'. By stating the concept in this way, Leroy did not contradict his idea that reasoning in carnivores differed from that of frugivores, but he considered the complexity of the cognitive processes of a species according to the 'objects' on which it operated. This may not be very accurate, of course, because many primates are fruit-eaters and their cognitive abilities are among the most complex in the animal kingdom. Leroy, a century before Darwin, also considered the many similarities between the behaviour of the few species he studied, and that of humans, while pointing out fundamental differences.

The writings of Aristotle, Cureau de la Chambre and Leroy show us that we can only get to know animals by observing them carefully and not by reading about them second or even third hand in books. They also show that what we know about animal behaviour comes from only a small sample of the millions of species that exist on our planet. Furthermore, this knowledge is immensely fragmented because of the small sample sizes of the populations that are usually studied. So, it is important to realize that when we say that we are speaking 'in the name of animals', it is highly pretentious and ignores the fact that we are talking about only a few 'select' species instead of the immense diversity of the animal kingdom, which is itself a part of an even greater biodiversity than plants.

A major turning point in our understanding of animals came when Darwin formalized and synthesized many of the ideas from the publications of animal observers before and after Descartes (Darwin, 1859). He extended the ideas of transformism and evolution and adopted the Aristotelian axiom of continuity between all species. This axiom had already been raised by Leibniz (1646–1716) in 1704 and then Linnaeus (1707–1778) in 1751 (p. 27) recalled that '*Natura non facit saltum*' (Latin for 'nature does not make jumps') which tells us that nature does not evolve in great leaps. This gradual continuity is evident in both the anatomical structure of animals and their behaviour. Nevertheless, despite his conviction about continuity, Darwin acknowledged that humans have adapted to an extent that has created a huge gap between them and anthropoids, the species most close to humans.

Following on from Darwin's theory of evolution, other scientists began to use experimental scientific approaches to study animal behaviour although it must be said that they still only considered very small samples of the huge diversity of the animal kingdom that surrounded them. Then, Romanes (1883) further defined Darwin's 'continuism' about the intelligence of species in his work *Animal Intelligence*, in which he provided a broad overview of species, from gastropods to primates. His approach was to resort to anecdotes, most of which were second or third hand. These anecdotes were certainly not scientific data, but were largely factual and therefore potential starting points for future scientific investigation. Romanes' interpretations of these anecdotes, while questionable, were a powerful stimulus for critical thinking, an essential element of the scientific approach.

The scientific approach to studying the behaviour of animals was taken up and advocated strongly by Morgan (1894). Like any biologist, Morgan observed the behaviour of animals, including his own dog but, like Romanes, he was particularly interested in how animals solved problems. However, unlike Romanes, he did not jump from observation to interpretation. Instead, he formulated hypotheses from his observations and tested these hypotheses on animals. Morgan made it very clear that 'the result of a two-minute incidental observation' differed greatly from the statistical results from a series of tests. His scientific approach was the foundation of comparative psychology and placed the study of behaviour, or ethology, firmly as a branch of the scientific field of biology. Thus, observation of what an animal does, the evolutionary comparative approach and the scientific approach were the three pillars that led to the emergence of ethology.

None the less, the dichotomy between naturalists and experimenters persisted at the dawn of ethology. Tinbergen (1952, 1953) and Lorenz (1957) elaborated the foundations of ethology within the framework of

Darwin's theories, particularly that of a continuity of behavioural expression between species. The key components were: (i) closely-related species express similar behavioural repertoires; (ii) species that are distant in origin express different repertoires; and (iii) behaviour can adapt because it is the expression of natural selection, which eliminates deleterious mutations and promotes mutations that benefit the species. Lorenz and Tinbergen's approaches postulate that there are, on the one hand 'innate' triggering stimuli, specific to each species or group of species (innate releasing mechanisms or IRMs), and on the other, stereotypical action patterns (fixed action patterns or FAPs).

This 'adaptive' vision of behaviour therefore inferred that behaviour was simply a product of evolution and so left little room for the concept of learning which became the main thrust of 'behaviourists' including Skinner (1953). None the less, Tinbergen noted that individual differences among animals must have resulted from some learning. Moreover, in his true manifesto of ethology, Tinbergen (1963) emphasized the importance of studying the development of behaviour, and by doing so, added a new dimension to the three fundamental questions that Huxley (1942) raised about the biology of behaviour: (i) the causality of behaviour; (ii) the immediate function of behaviour; and (iii) the 'ultimate' function of behaviour in the context of perpetuating the species. It was also in this publication that Tinbergen defined ethology as the 'biology of behaviour', thus clearly placing it as a scientific discipline.

Observing the behaviour of closely-related species has allowed, and continues to allow, ethologists to understand better the emergence and fixation of specific types of behaviour. This comparative approach remains an essential and fruitful tool in ethological studies. Tinbergen was originally a naturalist who observed individuals, both vertebrate and invertebrate, in their natural environment. His early ethological studies published in 1963 were largely naturalistic descriptions. However, he did use experimental approaches to test hypotheses on adaptive behaviours. But these experiments were far less rigorous than the restrictive approaches of the school of 'behaviourists', and of American ethologists trained in psychology (such as Lehrman, 1953, and Schneirla, 1954). In fact, experimental ethology was not widely practised until after the 1960s.

This change was partly the result of a meeting in 1954, between European ethologists, influenced by Tinbergen and Lorenz, and American researchers such as Schneirla (1902–1968) and Lehrman (1902–1972). Schneirla and Lehrman had been influenced by Morgan, the founder of comparative psychology, and were trained in psychology and experimental approaches to animal behaviour. This meeting fundamentally destroyed the concept of a clash between 'innate' and 'acquired', and the stalemate between Lorenz and Tinbergen on the one hand and behaviourists like Skinner (1904–1990) on the other. Lorenz and Tinbergen were proponents of Darwin's continuity and considered that behaviour was controlled solely by genes. Behaviour resulting from mutations that fixed specific differences was considered innate. As a result, they considered behaviour to vary little and, above all, not be subject to modification by apprenticeship and learning. In contrast, Skinner proposed that behaviour in animals and humans was controlled through learning throughout their lives and that phylogenetic constraints were negligible. The landmark confrontation of 1954 oriented ethological thought towards a fundamental interaction between an individual, with all its phylogenetic, anatomical and physiological constraints, and the environment in which it is placed. This position meant that phylogenetic constraints and the power of adaptability were important, and learning allows for this. This interactionist approach meant that we could account for individuals differing in behaviour, and recognize the power of adaptive changes during the development of the individual, or ontogenesis. This brings us to the fourth question that Tinbergen (1963) wished to add to the three other fundamental questions of biology proposed by Huxley – that of the development of behaviour.

Ethologists and comparative psychologists have accumulated quantitative, reliable, scientific evidence from observing samples of individuals that should be as large as possible. This evidence, obtained from '**data**', is far stronger than empirical information. When studying behaviour, empirical information and scientific information overlap. Sometimes they oppose, and sometimes they complement one another. Both have their origins in fact. Empirical information sticks to these facts and is characterized by its non-probabilistic, non-analytical and, above all, non-reliable character. It is based on empirical observations and intuitions and purports to be efficient and pragmatic. This approach generates among readers the belief that they are difficult to debate because they come only from the authors' intuitions and are therefore difficult to share. By contrast, scientific knowledge is analytical. The observed fact becomes a hypothesis that, after a long process, leads to a scientific result. This information is probabilistic because it comes from the analysis of the behaviour of many individuals. All stages of the process – explanation of hypotheses, methods, and analyses of cause-and-effect relationships – are arguable. Allowing everyone to take ownership of the conclusions of scientific work means that all stages of this work are accessible and explained.

A small example illustrates the gulf between the intuitive, empirical interpretation of practices, and the results from well-reasoned hypotheses and scientific knowledge. In the 1990s, farmers placed heifers from a suckling herd in an enclosure and tethered them at the time of weaning to deworm them. They noticed that their animals were unusually docile afterwards and attributed this docility to the restraining and close association between the heifers and the farmers. A scientific analysis, exploring the effects of positive contacts at different periods of heifers' lives on their subsequent relationship with humans, showed that contact with humans just after weaning, established a positive relationship that persisted for the life of the animal (Boivin *et al.*, 1992). This work showed that positive contacts at a specific moment generated a lasting acceptance of the human by the heifers, and not the tethering and coercion. This principal was later confirmed in other domestic species. The scientific approach made it possible to avoid maintaining an intuitive belief and lead to proposals of more general practices based on reliable analyses.

Unfortunately, in ethology, as in all sciences, there are some publications or other forms of dissemination of information in which claims are made as a result of scientific analyses, but because of the weakness of their hypotheses, the methods used, or the inconsistency of the conclusions, these studies ought not be accepted for publication in peer-reviewed scientific journals. In some cases, the authors justify not going through the peer-review stage by arguing the urgency of 'getting the information out there'. These 'pseudo-scientific results' lead to the establishment of beliefs rather than principles. Those who produce them appear more like gurus than scientists, but not in the eyes of the general public who cannot exercise their critical mind on these 'irrefutable' claims.

1.2   A Science that is Constantly Developing

Ethologists are constantly seeking answers to questions from many species in many environmental conditions. Some confine their studies to the 'natural' environments in which the species has evolved. Others use environments that are very different from the natural environment, so that they can use experimental approaches that allow variables to be controlled better. With domestic species that have commercial value they study them in environments that are more or less those of conventional animal husbandry. By contrast, companion animals often pose difficult problems in methodology because of their complex relationships with humans. Whatever the method of study, answers to the experimental questions posed provide information either about the fundamental biology of the species studied or about issues stemming from 'modes' of current thought. Despite this, we have so far gained only small snippets of information on how the lives of only a very few animal species relate to ours. And this information is more often about domestic rather than exotic species. So, our knowledge of the complexity and diverse nature of the behaviour of animals is like a vast piece of lace, with very elaborate patterns but with an even larger number of holes.

In fact, the development of ideas and knowledge about behaviour has much in common with the concept of biological evolution put forward by Darwin. He proposed that permanent mutations in biological material lead to changes in how species adapt. When these adaptations are favourable, they spread rapidly within populations because of the success they provide. When they are neutral, they remain dormant but, when they are deleterious, the individuals carrying these mutations disappear.

Like genetic material, observations about behaviour in animals are also constantly moving and changing so that new conceptual frameworks, even theories, are proposed from time to time. Some have become successful and generate a lot of research activity. But, over time, these concepts may become diluted or even disappear when other concepts emerge from the same or new information or from previous results. There is a permanent activity between creativity, refutation, forgetfulness and resurgence.

Behaviour is central to the study of how animals relate to one another. It can only be approached by careful observation of what animals do when faced with different environmental challenges and the scientific data that they generate. Sometimes these data are confronting and sometimes they are poorly presented in the literature so that, sadly, the work of researchers, ethologists or comparative psychologists, is too often ignored, decried or even rejected. For example, Scott and Fuller (1965), authors of many original papers on dogs and a critical review of their genetics and social behaviour, reported that it was difficult to do research on dogs because everyone 'knows' all about them. Nothing has changed. In fact, difficulty has probably increased, despite a recent wave of scientific work, maybe too much, on the behaviour of dogs. Much of this work followed a naturalistic approach and has led to high-quality empirical knowledge. Perhaps a scientific approach might have led to challengeable knowledge and the use of statistics to account for individual variability and other valuable information. As Firestein (2012, p. 10) points out, 'scientific knowledge generates a cultured ignorance of superior quality'.

In short, we can conclude that at this stage in its development, ethology, or the biology of behaviour, analyses the interaction between individuals and the environment in which they find themselves. This environment can be the natural one in which they evolved without human interference. It can also be a captive environment for individuals of a wild species, or it can be anthropized in the case of domestic species.

1.3 Ethologists – Researchers Who Work With the Subjects They Study

Ethologists gather information on the species they study by observing the way they live. In doing so they are following the counsel of Tinbergen (1953, p. 139), 'I must emphasize that thorough reading, while necessary, will never replace "first-hand" knowledge based on personal observations. The animals themselves are always more important than the books that have been written about them.' He claimed that ethology had become a quantitative discipline and its place as a behavioural biology was fully justified. In other words, ethologists don't just observe behaviour; they do it according to the concepts of scientific discipline. Observation is only the stage when data are collected. These data must then be analysed as part of the scientific method to generate a hypothesis. An integral part of this process is to choose carefully and define precisely the behavioural variables that are collected from individuals so that they can be quantified. In this form they make it possible to accept or reject hypotheses put forward at the outset. As a simple example, let us suppose the question asked concerns the study of the relationship between speed of movement of a horse and its age. It would be preferable to define behavioural units such as 'walking', 'cantering', 'trotting', 'galloping' or 'jumping', and not use a more general unit such as 'moving' which would include all of these.

After the repertoire has been decided, variables have to be collected according to a sampling plan. This sampling must take into account the rhythms of the individuals and relate clearly to the hypothesis, or the question asked. A fundamental characteristic of ethology is that it is applied to individuals. Even if the ultimate unit is a group, each individual in this group should be taken into account (Altmann, 1974) so that the researcher can calculate the variability between those individuals. From this can be calculated the 'background noise', which in turn determines how precise or how general one can be when making conclusions about the phenomena studied and the links between individuals. Thus, in the example of the horses above, even though individuals among the young animals may be lethargic and some of the older adults may be very active, a hypothesis may still be sustained to verify that young subjects are more active and move in a more varied and rapid way than adults.

Depending on the study, researchers may decide on the methods they use based on the context in which animals usually live. This form of research is described by Altmann (1974) as 'naturalistic'. Alternatively, the methodology can involve measuring variables in an environment that has been deliberately modified – in other words is 'experimental'. In both cases, and no matter how the data are collected, the ethologists plan their observations to match the way their animals act. Those animals continue to depend on their biological rhythms and ethologists must study behaviour throughout the cycle of this rhythm. For example, animals sampled only in the middle of the day would be unlikely to manifest all of the distinctive patterns of behaviour they would normally show during a whole day. On the other hand, diurnal species are probably only active between sunrise and sunset so there may be little sense in ethologists continuing to study them throughout the night. In addition, ethologists must avoid disturbing the 'background' environment of the animal. Such disturbance could affect the behaviour of animals resulting in temporary disturbance, or immobility, fear (freezing), or even worse, total flight. Thinking about the method being used is also important when organizing the way in which data are collected. This is because, ultimately, researchers have to draw conclusions from the data they collect, and the right methodology can make that task much easier. Approaching scientific protocols in this way is sometimes made light of, even by some scientists. But, especially for the ethologist, who works in a field where a lot of information may appear to be contradictory, meticulous data collection is essential. Such data offer the only rational way to analyse, criticize or possibly refute theories about behaviour. Mere observations or more and more empirical information, even though they may sound attractive, cannot match the approach of the serious ethologist.

1.4 Describing Animal Behaviour

Concepts that are developed from studies of animal behaviour are ultimately described in words. The choice of these words may often be critical when trying to express concepts about behaviour and mental states clearly. Words are chosen by humans and are intended for humans. Even humans are happy to use words that cover experiences that we cannot directly feel. For example, giving birth or breastfeeding are activities that only women experience. None the less, men, who never have direct access to it, talk about it in the same way as women. Similarly, orgasm in both men and women is a personal experience that is not very shareable. However, we do have words for these experiences. This is not a serious problem when we are translating the actions of humans because we have an intuitive knowledge of their meaning, even if this intuition is sometimes misleading. On the other hand, it may be a major problem when it comes to describing non-human species.

Because of this, some people believe that the vocabulary ethologists use to describe animals should differ from that which they might use to describe humans. If they are right, then why should not each animal species have its own special vocabulary? Behaviourists like Skinner, used to create words to describe the situations they observed. These words were similar to those which one might use to describe how robotic systems work and they were very descriptive. It may well have been their way to distance their own feelings from the activities they observed in animals, but Lorenz noted that it was not possible to interrogate animals effectively without having an understanding based on love and respect (de Waal, 2016, p. 19). De Waal illustrated this by using the example, 'reconciliation sealed by a kiss'. He concedes that the phrase could have been written, 'post-agonist interaction with oral–oral contact' (Camos *et al.*, 2009), but then pointed out that the meaning of the phrase would probably be lost! It seems to us that Skinner's rigid objectivity introduces a sterile bias because, 'we have to remember that what we observe is not nature in itself but nature exposed to our method of questioning' (Heisenberg, 1958, p. 58).

We believe that the most useful approach is to start with the idea that there is no difference in nature between humans and animals, but a difference in level that we can test. So, we can speak of aggressiveness in both humans and animals, but also of awareness, expectations, fear, suffering, pain, pleasure, using the same words as in humans with the proviso that these words are defined. This would certainly be better than introducing words unsupported by a clear definition and therefore not challengeable. It should also be borne in mind that the absence of evidence is not evidence of absence (de Waal, 2016, p. 13).

If we use this approach, words need not be specific to a species, but they should be defined in the context of the study. The operational definition must be able to be challenged and tested by experiment. It can still be questioned if the results do not support the initial hypotheses (Lindsay, 2020). To illustrate, let us look at 'episodic memory'. For a long time, it was believed that episodic memory was confined to humans because it required verbal answers to questions asked. So, for animals the hypothesis was not tested because animals could not answer verbally. However, Clayton and Dickinson (1998) developed experimental devices to test episodic memory in the western scrub jay by inducing these birds to respond non-verbally to stimuli. This appeared to overcome the problem. There were still some critics of this work but Suddendorf and Corballis (2010) conducted new experiments using the original experimental technique and had the same results. Most ethologists now believe that episodic memory exists, at least in some species, if we can devise ways to measure it. In the case of the western scrub jay, the technique of measurement was totally opportunistic. The researchers noticed that these birds hid their food – acorns or other seeds – in the autumn and recovered them several months later when food was lacking. Their capacity to do this successfully suggested that they had memorized the location of the caches when prompted by changes in the season. Recovery of food was therefore a substitute for a verbal response when studying memory. We haven't yet studied episodic memory in many other species, probably because we have not been creative or intelligent enough to identify similar proxies to replace language (de Waal, 2016). Perhaps this also applies to cognitive abilities other than episodic memory.

The creation of a specific vocabulary for animals would de facto lead to a notional barrier between humans and animals. To use the term, 'animal understanding' would suggest that human understanding is different from animal understanding, and this seems counterproductive. Moreover, we have no idea which animals this definition might apply to. So, a separate vocabulary for each species or each population would make futile efforts by experimenters to generalize.

We also stress the profound difference between a word and a concept. A concept is obviously designated by a word, its label. But words can be replaced by synonyms, while a concept can only be described in relation to other concepts. Concepts are based on properties that are unique when compared to all other concepts. Often, we see a concept used as a simple word, and therefore distorted. Fortunately, ethologists make little use of 'scholarly words', or newly coined words made from Greek roots to make them sound 'academic' – such as neologisms! Descriptions of animals, especially those in the wild, are often embellished with judgements about the 'intentions' of animals. Intentions are complex mental processes, and it is debatable whether they can they be attributed to animals.

1.5 Domestic Animals as Subjects of Study

All animals are the result of myriads of generations of selections in their original environments. We can consider that the state in which we find them after these countless generations is the one that allowed the population to maintain itself within the constraints of the environment. This balance can be called the 'natural' state, although it may not necessarily be absolutely optimal. It is just the best the animals can do in their environment. Domestic animals are that tiny fraction of the world's animal species that has passed their most recent generations in environments in which humans have had an influence.

The Farm Animal Welfare Council (FAWC) (1979) describes five freedoms for domestic animals. The first four of those freedoms can be considered negative: (i) freedom from hunger and thirst; (ii) freedom from discomfort; (iii) freedom from pain, injury and disease; and (iv) freedom from fear and distress. The fifth is positive: freedom to express normal and natural behaviour (e.g. providing roosts for chickens to satisfy a natural instinct to roost). The World Organisation for Animal Health (OIE) and the European Commission have adopted a similar framework that focuses on 'the freedom to display most normal patterns of behaviour'. This idea is not easy to put into operation because it is difficult to define what constitutes 'natural behaviour of the species', especially when we consider individuals in populations of domestic animals. These species have almost certainly undergone artificial selection to a greater or lesser degree during their long association with humans. Breeders will have consciously or unconsciously selected them as breeding stock based on physiological or behavioural criteria that they considered desirable. Breeders certainly control the husbandry of the animals but the most important driver for adapting animals to the needs of animal industries is selection.

Not all selection is straight forward. Some traits, including milk production, can only be selected directly in females but the choice of males for breeding is more critical for success since a bull can have many more offspring than a cow. Genetic progress is therefore often faster when breeders select males that sire suitable females than when they only select females. A similar, off-beat, example is that of fighting bulls whose 'value' in the arena is not known until after they die. It was assumed that the substitute criteria for selection were as close as possible to those of the desired offspring. In some strains of poultry, pecking behaviour can be a problem in intensive farming (Blokhuis and Wiepkema, 1998; Cronin *et al.*, 2018). Pecking is normal when it is limited to using the beak to pick up pellets or non-food items. But especially in confinement and at high population densities, some individuals may systematically peck their companions leading to injury and death. Breeding the surviving animals actually increases the frequency of pecking in the population since the surviving animals are those that peck. And pecking is a trait that is genetically controlled (Rodenburg *et al.*, 2008).

In domestication, selection focuses on traits for productivity, but behavioural components of these traits are also known to be in play, as the examples of laying hens and dairy cows illustrate. However, we highlight here the 'hypertypes', or those that accentuate extreme of distinctive behavioural features in specific animal populations. In laying hens, breeders have selected for egg production for many generations. This has resulted in limiting or eliminating brooding behaviour altogether. Broodiness is, of course, essential in birds where humans do not intervene, but in intensive farming, brooding and caring for the young, interrupts egg production. Artificial brooding has replaced this phase entirely on commercial egg farms. These days, only ancient breeds of hens, mostly kept by collectors and enthusiasts, have retained the ability to become broody. Similar distortion of natural behaviour is seen in commercial turkey raising. Males of modern strains have become so large due to selection for weight that they cannot mate females. Therefore, reproduction is only possible by artificial insemination.

In some cases, productivity is not the main objective and in some breeds of cattle, behavioural traits rather than productivity are selected. In most dairy cows for example, farmers favour cows that are docile towards both humans and each other. This is not considered so important for bulls. In other species, breeders have selected aggressive animals. Breeders of roosters or dogs that are bred for fighting seek animals that behave aggressively. Also, aggressiveness towards humans is preferred in the fighting bulls used for the Spanish and Camargue arenas.

Even when not explicitly considered in selection programmes, behavioural traits sometimes intervene unpredictably. In milk-producing cattle, especially Holsteins, intense selection for milk production has been accompanied by an increase in lameness and mastitis and a decrease in their reproductive efficiency (Chagas *et al.*, 2007). Calves from dairy breeds such as Holsteins which are weaned at a few days of age learn quickly to drink from a bucket, but those from beef breeds, where they are normally suckled for many months, do not. Dairy cows readily adapt to being milked and have a milk ejection reflex even when their calves are absent. They also rapidly accept fostering of foreign calves. On the other hand, suckler cows are very selective, and allow only the calves they adopted to receive milk. Attempts by humans, machines or foreign calves to take milk result in their rapidly ceasing lactation altogether. This is common in many other species, including horses, yaks, zebu and camels, in which females dry up quickly if they are not sucked by their own young.

Many studies over the last 30 years have sought to define an optimal environment for domestic animals or single out the most important elements of their environment. Now we have information on such things as cage sizes for hens and veal calves or suitable floor surfaces for housing broilers. These studies have also tested ways of modifying the environment, with the aim of 'enriching' it and defining the elements of the environment that permit animals to behave normally. These tests acknowledge that, in fact, it is not the environment that one wishes to enrich but the behaviour of the animals experiencing that environment. Changing the environment can often lead to positive changes in the behaviour becoming more diverse. In this case, and only in this case, can one conclude that there is 'enrichment' of the environment.

Breeders and researchers have sought solutions to problems arising when animals are constrained. Among these solutions, some of which are 'radical', are various mutilations including the castration of pigs, cattle and sheep, the trimming of teeth and docking tails of piglets, and the debeaking of laying hens. These mutilations raise ethical problems, since the solution chosen to eliminate the behavioural problem is to injure the animal instead of modifying husbandry practices that may have generated the behavioural problem in the first place.

One ethical solution might be through selection: for example, breeding pigs without teeth or without tails – so-called 'manufacture by artificial selection'. This avenue has been explored, including selection against lameness of strains of broiler hens with high growth potential, and selection against mastitis and lameness in dairy cows with high potential for production. These practices are only used to limit the appearance of anomalies when animals are handicapped by systems in which the main objective is productivity. They are often only palliatives to constraints that were imposed by the breeder in the first place.

To define conditions of husbandry that ought to be retained or avoided, some researchers have studied the reactions of animals in farming conditions rich in resources and with the minimum of constraints. For example, Stolba and Wood-Gush (1989) observed pigs that lived in groups with free access to the outdoors that were only disturbed for health care and feeding. This was far from natural breeding in the wild where animals would have had to look for food and shelter, and fight against predators and parasites. The same approach was adopted for chickens and cattle. Nowadays significant numbers of broiler chickens are raised for at least part of their lives outdoors. Similarly, most veal calves nowadays live in groups on straw bedding instead of stalls. These experimental analyses avoided using the pitfall of evaluating environmental conditions that suited the farmers and not the animals and addressed the issue not of whether we are satisfied with the living conditions of the animals, but whether the animals are satisfied with the conditions imposed on them.

Yet other domestic species, especially dogs and cats, are artificially selected for criteria that are purely aesthetic. In dogs, the combination of aesthetic and utilitarian criteria has led to the creation of hypertypes where the criteria appear to be pushed to the extreme, such as prognathism, or distorted lower jaw, in bulldogs, or the excessively wrinkled skin in the Shar Pei breed, or dwarfism or large size in several other breeds. These hypertypes have become associated systematically with various pathologies that require the intervention of humans for treatment.

Animals of populations that live without constraint imposed by humans have behaviour that is the result of natural selection. However, our concept of naturalness of behaviour is often a fantasy related to a vision of a world, in which the 'wild' environment without humans is considered to be an optimal situation and far superior to anything that humans could produce. A wild species is a product of natural selection leading to its adapting over time to a given environment. Therefore, that environment is the best reference to observe what is happening within its network of limitations. It is also obvious that the domestic environment is not natural and that it alters considerably the way an animal lives. So, domestication opened up entirely new constraints and possibilities by imposing new living environments that have, in turn, developed into new criteria for which we should select. So, constantly pushing the constraints we impose on animals to extremes cannot be done without consequences on their quality of life. It has been suggested by Larrère (1996) that we not impose constraints on animals that they would not be able to face without human interference. In other words, a realistic approach would be to check that conditions imposed on animals do not exceed their capacity to adapt, do not generate suffering, and allow animals to live in comfort. Such an environment would have as few constraints as possible (Broom, 2021).

References

Altmann, J. (1974) Observational study of behavior: sampling methods. *Behaviour* 49(3), 227–267. DOI: 10.1163/156853974x00534.

Blokhuis, H.J. and Wiepkema, P.R. (1998) Studies of feather pecking in poultry. *The Veterinary Quarterly* 20(1), 6–9. DOI: 10.1080/01652176.1998.9694825.

Boivin, X., Neindre, P.L. and Chupin, J.M. (1992) Establishment of cattle-human relationships. *Applied Animal Behaviour Science* 32(4), 325–335. DOI: 10.1016/S0168-1591(05)80025-5.

Broom, D. (2021) *Broom and Fraser's Domestic Animal Behaviour and Welfare*, 6th edn. CAB International, Wallingford, UK, 573 pp.

Camos, V., Cézilly, F., Guenancia, P. and Sylvestre, J.P. (2009) *Homme et Animal, La Question Des Frontières*. Éditions Quae, Versailles, France. DOI: 10.35690/978-2-7592-0323-9.

Chagas, L.M., Bass, J.J., Blache, D., Burke, C.R., Kay, J.K. *et al.* (2007) Invited review: new perspectives on the roles of nutrition and metabolic priorities in the subfertility of high-producing dairy cows. *Journal of Dairy Science* 90(9), 4022–4032. DOI: 10.3168/jds.2006-852.

Chanet, P. (1646) *De l'Instinct et de la Connoissance des Animaux*. Toussaincts de Govy, La Rochelle, France.

Clayton, N.S. and Dickinson, A. (1998) Episodic-like memory during cache recovery by scrub jays. *Nature* 395(6699), 272–274. DOI: 10.1038/26216.

Cronin, G.M., Hopcroft, R.L., Groves, P.J., Hall, E.J.S., Phalen, D.N. *et al.* (2018) Why did severe feather pecking and cannibalism outbreaks occur? An unintended case study while investigating the effects of forage and stress on pullets during rearing. *Poultry Science* 97(5), 1484–1502. DOI: 10.3382/ps/pey022.

Cureau de la Chambre, M. (1647) *Traité de la Connoissance des Animaux, Où Tout Ce Qui a Esté Dict Pour et Contre le Raisonnement des Bestes est Examine*. Pierre Rocolet, Paris, 390 pp.

Darwin, C. (1859) *On the Origin of Species by Means of Natural Selection, or, The Preservation of Favoured Races in the Struggle for Life*. John Murray, London. DOI: 10.5962/bhl.title.82303.

Descartes, R. (1637) *[1982] Discours de la Méthod*, 10/18. Librairie Philosophique, J. Vrin, Paris.

de Waal, F. (2016) *Are We Smart Enough to Know How Smart Animals Are?* Granta, London, 340 pp.

Farm Animal Welfare Council (1979) Farm Animal Welfare Council Press statement, 5 December. Available at: http://webarchive.nationalarchives.gov.uk/2 0121007104210/http:/www.fawc.org.uk/pdf/fivefreedoms1979.pdf (accessed 10 May 2022).

Firestein, S. (2012) What science wants to know. *Scientific American* 306(4), 10. DOI: 10.1038/scientificamerican0412-10.

Heisenberg, W. (1958) *Physics and Philosophy: The Revolution in Modern Science*. Harper and Brothers, New York, 207 pp.

Huxley, J.S. (1942) *Evolution: The Modern Synthesis*. Allen and Unwin, London.

Larrère, R. (1996) Le contrat domestique. *Courrier de l'Environnement* 30, 1–17.

Lehrman, D.S. (1953) A critique of Konrad Lorenz's theory of instinctive behavior. *The Quarterly Review of Biology* 28(4), 337–363. DOI: 10.1086/399858.

Leibniz, G.W. (1704) *New Essays on Human Understanding*, Second edition translated and edited by Remnant, P. and Bennet, J. Cambridge University Press, New York.

Leroy, C.G. (1768) *[1896] Lettres sur les Animaux*, 5th edn. Éditions Vigot, Paris, 264 pp.

Lindsay, D.R.L. (2020) *Scientific Writing = Thinking in Words*. CSIRO Publishing, Clayton, Victoria, Australia. DOI: 10.1071/9781486311484.

Linnaeus, C. (1751) *Philosophia Botanica*, 1st edn. Kiesewetter, Stockholm.

Lorenz, K. (1957) *Essais sur le Comportement Animal et Humain*. Seuil, Paris, p. 477.

Louis, P. (1969) *Aristote, Histoire des Animaux, Textes Établis et Traduits par Pierre Louis*, Vol. 1 and 2. Denoël, Paris.

Marler, P. and Hamilton, W.D. (1966) *Mechanisms of Animal Behaviour*. Wiley, New York, 771 pp.

Morgan, C.L. (1894) *Introduction to Comparative Psychology*. W. Scott, London. DOI: 10.1037/11344-000.

Rodenburg, T.B., Komen, H., Ellen, E.D., Uitdehaag, K.A. and van Arendonk, J.A.M. (2008) Selection method and early-life history affect behavioural development, feather pecking and cannibalism in laying hens: a review. *Applied Animal Behaviour Science* 110(3–4), 217–228. DOI: 10.1016/j.applanim.2007.09.009.

Romanes, G.J. (1883) *Animal Intelligence.* Appleton and Co, New York. DOI: 10.5962/bhl.title.1046.

Schneirla, T.C. (1954) Interrelationships of the 'innate' and the 'acquired' in instinctive behavior. In: Grassé, P.P. (ed.) *L'instinct dans le Comportement des Animaux et de l'Homme.* Masson, Paris, pp. 401–442.

Scott, J.P. and Fuller, J.L. (1965) *Genetics and the Social Behavior of the Dog.* University of Chicago Press, Chicago, Illinois.

Skinner, B.F. (1953) *Science and Human Behavior.* Macmillan, London, 461 pp.

Stolba, A. and Wood-Gush, D.G.M. (1989) The behaviour of pigs in a semi-natural environment. *Animal Science* 48(2), 419–425. DOI: 10.1017/S0003356100040411.

Suddendorf, T. and Corballis, M.C. (2010) Behavioural evidence for mental time travel in nonhuman animals. *Behavioural Brain Research* 215(2), 292–298. DOI: 10.1016/j.bbr.2009.11.044.

Tinbergen, N. (1952) Derived activities: their causation, biological significance, origin. *The Quarterly Review of Biology* 27, 1–32.

Tinbergen, N. (1953) *Social Behaviour in Animals.* Methuen and Co, London, 188 pp.

Tinbergen, N. (1963) On aims and methods of Ethology. *Zeitschrift Für Tierpsychologie* 20(4), 410–433. DOI: 10.1111/j.1439-0310.1963.tb01161.x.

The Senses: How Do Animals Know What Is Happening to Them?

2.1 Sensory Perceptions: Show Me Your Brain and I'll Tell You Who You Are!

The five senses, touch, vision, hearing, olfaction and taste, allow individuals to perceive the permanent or even fleeting stimuli in the environment around them. They can select those stimuli that are relevant to their needs at a given time. For instance, when the concentration of circulating sex hormones is high the relevant stimuli are different from when they are not. These stimuli encourage individuals to target individuals of the same species, to consider copulation and, in some cases, to build lasting relationships.

To study the importance of the senses of a species we need to examine the region in which most of the sensory receptors are concentrated – the area of the brain known as the sensory cortex (Krubitzer, 1995, 2007; Krubitzer and Kaas, 2005; Krubitzer and Seelke, 2012; Krubitzer and Stolzenberg, 2014). Comparing the brain structures of different mammalian species shows that the same structures are organized very similarly regardless of the species, and that these structures vary in volume depending on the species and especially the way they have adapted. In Fig. 2.1 we can see that the auditory cortex of bats is highly developed compared with that of squirrels, while the visual cortex in squirrels is much more developed than that of bats. Bats move by echolocation, emitting ultrasound and responding to its echo which enables them to move at night and detect and capture flying prey. The capacity of bats to do this is readily demonstrated when we observe their auditory cortex and how much its volume has developed during evolution. We see similar evidence of specialist parts of the sensory cortex being selectively developed when looking at animals that depend heavily on motor activity or specific anatomical features when seeking food. For instance, the nose of the platypus is vital to its search for food. It is richly innervated, and the endings of these nerves appear in its somaesthetic cortex, which resembles the shape of the platypus's nose. A

© CAB International 2022. *Interacting with Animals: Understanding their Behaviour and Welfare* (Pierre Le Neindre and Bertrand L. Deputte) DOI: 10.1079/9781800622418.0002

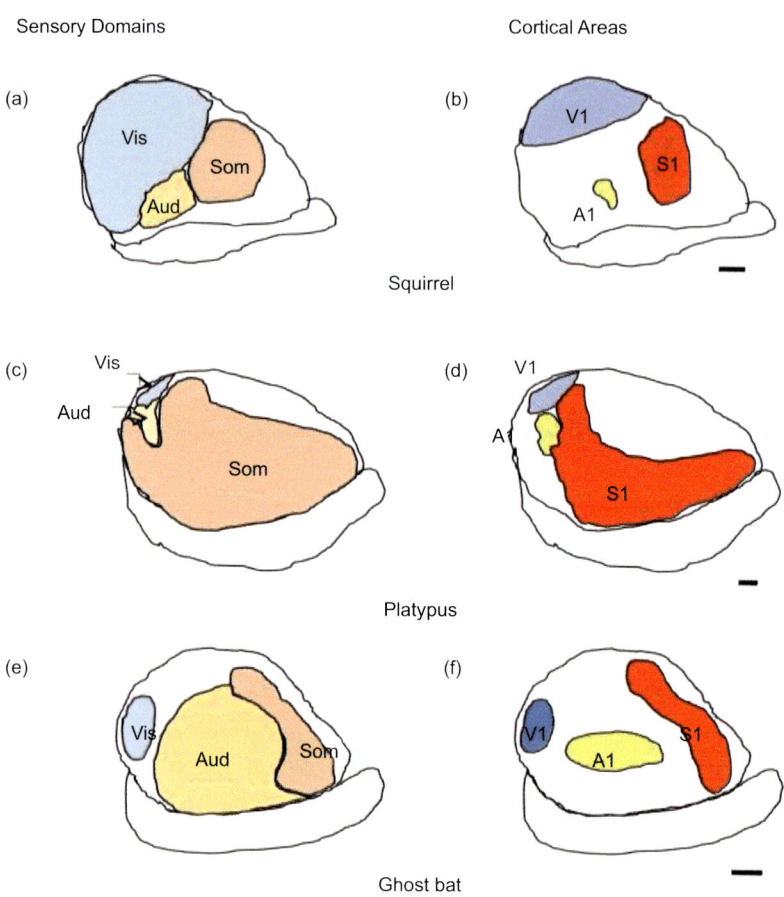

Fig. 2.1. Allocation of sensory domains – vision, audition, somaesthetic – in the cortex of three different mammal species living in contrasting environments. Sensory domains are presented in the left column, the cortical fields in the right one. For the different species, while the localization of the cortical areas is maintained, their sizes differ according to the ecological adaptations of the species. (a and b) The squirrel, an arboreal species, has a developed vision and the largest part of its neocortex is devoted to the visual system (Vis and V1 in the left and right column, respectively). (c and d) The platypus or duck-billed platypus (*Ornithorhynchus anatinus*) has a highly developed bill. It lives underwater where it captures prey. Its bill contains a high density of mechanoreceptors and electrosensory receptors. The somaesthetic – somatosensory – cortical domain is the most developed part of the brain (Som and S1 in the left and right column, respectively). (e and f) The ghost bat (*Macroderma gigas*) is a mammal using echolocation to perform its most important behaviours. The largest part of its neocortex is devoted to the auditory system (Aud and A1 in the left and right column, respectively). Scale bars = 1 mm. Redrawn and adapted from Krubitzer and Kaas, 2005.

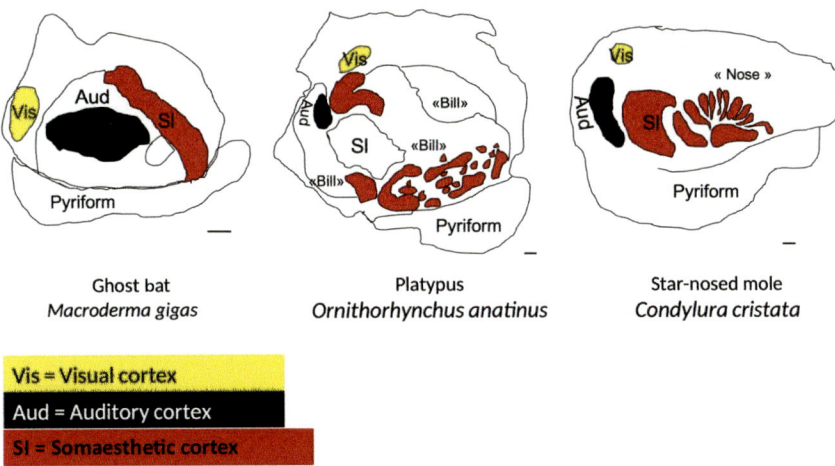

Ghost bat
Macroderma gigas

Platypus
Ornithorhynchus anatinus

Star-nosed mole
Condylura cristata

Vis = Visual cortex

Aud = Auditory cortex

SI = Somaesthetic cortex

Fig. 2.2. Organization of the neocortex in species that display very special adaptations. Subdivisions of the cortex in species with particular sensory specializations, such as echolocation in ghost bat (*M. gigas*), electroreception in the platypus (*O. anatinus*) and somaesthesis in the star-nosed mole (*Condylura cristata*). More than half of the cortex of the ghost bat is devoted to the treatment of auditory information (black+ Aud). The somaesthetic information, essential in the life of the platypus, is mainly processed in its cortex in area SI. The projections of the somaesthetic receptors form the shape of the platypus bill, above the pyriform area. This is also the case for the star-nosed mole which relies mostly on sensory information from its nose. The particular star shape of the nose is reflected in the projections of the sensory receptors on the cortex (somatotopy: « Nose » in the right part of the brain). The dorsal part of the brain is at the top and its rostral part to the right. Scale bars = 1 mm. Redrawn and adapted from Krubitzer, 1995.

similar and spectacular example is the star-nosed mole (Fig. 2.2). Its nose is also very important to the way it lives, and it too is heavily innervated and its morphology is reproduced in the sensory cortex (Catania, 2011).

However, to understand behaviour in vertebrate species we need to have information other than just their anatomy. Unfortunately, research into how animals perceive the world is extremely fragmentary especially when we consider the immense diversity of behavioural adaptations in the animal kingdom. Also, we are apt to analyse their behaviour in relation to our own human perception of the world. As a result, experimenters have developed techniques that are increasingly elaborate to explore in animals those senses that are also important to us, such as vision and hearing. But smell, in particular, is a sense for which we humans are poorly equipped compared to many species of non-primate mammals and which, in many cases, is very important to them. Moreover, the volatile nature of olfactory information makes investigating the role of smell very complex. So we must look at concentrations of olfactory stimuli that may seem to be very low for us when studying what an individual or group of individuals can perceive, before we can draw conclusions that are not ambiguous.

Similarly, when studying hearing in animal species researchers must also look at thresholds that are sometimes well outside the range of humans and primates. Like human audiologists who use audiograms to determine whether their patients need hearing aids, researchers in animals also use a continuum of sound frequencies (expressed in hertz or Hz) to decide thresholds below which sound is no longer perceived by animals and to which they are unlikely to respond. Again, animals can have a range that differs greatly from ours. We humans perceive sound frequencies in a range from around 31–17,600 kHz (Heffner and Heffner, 1983). But rodents, bats and dogs and many other species perceive ultrasound frequencies that are at least 2–4.5 times higher than that of the highest sounds that we humans can hear.

Also we need to study the diversity of stimuli that animals can perceive *and* discriminate against or consider in some way. For example, with vision, objects can be perceived during photophase, during the day, or scotophase, at night, and their shape and colour may vary according to the light. Most humans, apart from those who are colour blind or who have other visual problems, see the world as a palette of colours that we can name and recognize as many shades. As primates, our sensory equipment is that of trichromatic species – that is, we use three pigments to give us a rich spectrum of colour. However, many New World monkeys in South America are nocturnal and are dichromatic. In biology 'the exception is the rule'!

Birds perceive a much broader palette of colours than we do because they recognize a fourth pigment – ultraviolet – so their vision of the environment must be very different from ours and it must give them a set of values about stimuli that we are unable to recognize. In contrast, all non-primate mammals, including all domestic animals are dichromatic, so their colour vision of their environment is less rich than ours. For example, dogs do not see the colour red. Therefore, we must account for these dimensions, shapes and colours when seeking to understand behavioural reactions across species.

Spatially, stimuli have three dimensions, but in some experimental circumstances they can be presented in two dimensions, like a representation on a screen or in the form of a poster. Stimuli can also be static, and consequently easily discerned, or dynamic which makes investigating them much more complex. The receptors that respond to stimuli are also diverse. Tactile receptors are widely distributed throughout the skin. And the skin is the envelope that is in constant contact with the external environment. It allows the individual to feel heat, cold, humidity, wind direction and other, perhaps more subtle, differences in the characteristics of its milieu such as the possibility of passing through a narrow opening.

The position of the eyes on the face determines how well binocular vision can allow three-dimensional awareness of objects. Among the higher vertebrates, only birds and primates cover a wide visual field due to the position of both eyes in relation to their facial mass. Other mammals, such as cattle, horses and canids, usually have a developed facial configuration that reduces the extent to which the visual field of each eye overlaps. However, they usually compensate for this with an increase in visual fields to the side and back of the head.

This highlights the fact that each species can have unique sensory abilities that can have little or no relationship with other species and, certainly, they can be very different from those of humans.

2.2 How Do Animals Perceive Things?

Perception stems from the many methods of communication that are in play when two individuals of the same species interact. This multimodality tells us that the signals emitted, especially when emitted simultaneously, during an interaction are related to several different motor sensory channels. For example, a vocalization may be accompanied by a particular gesture or a posture, in other words it is bimodal – acoustic and visual. This is a common bimodality that can be used by researchers to determine how vocalizations work. They can study gestures made by the animal when it makes a sound. The result is much more precise than the sound alone, which is often too vague. Despite this, some authors have described this as a catch-all that indicates a lack of rigour in determining the essential elements and function of vocalizations (Menzel and Johnson, 1976, p. 131). For example, a vocalization from an animal associated with a stiff posture, stretching forward, and looking menacingly towards a member of the same species, might be described as an 'aggressive' vocalization. Another example of bimodality is when cats rub themselves on objects, including the legs of familiar humans. Rubbing is a tactile modality. When they sniff on the rubbed parts it is an olfactory modality. So rubbing by cats is a tactile *plus* olfactory bimodality. This is probably how cats familiarize themselves with their environment. They rub to 'absorb' olfactory information. During toileting, or self-grooming, they cover much of their own body with their own smell. But after episodes of exploration, feeding or interaction with humans, the cat can 'reset' its olfactory information.

2.3 Which Sense Organs?

All the examples above show that non-human animals do not perceive, in fact cannot perceive, the world as we do, or that they perceive it quite differently. They can capture information, visual, acoustic or olfactory, that our human sensory equipment does not allow us to discern. Each animal species has adapted to its own perceptual world, so we must be careful to remain aware of these adaptations and not to attribute our own specificities to all other species.

von Uexküll (1934) described the differences that unlike species perceive from their environment. In each species, and within species, there is a 'living environment', an *Umwelt*. He considered that this *Umwelt* had two parts: (i) a *Merkwelt*, a world of specific perceptions depending on the characteristics of the sense organs of the species; and (ii) a *Werkwelt*, a world of actions that determines the experience that each individual builds for itself. This notion of *Umwelt* underlined the diversity of adaptations among species, and therefore the differences between species. The concept is particularly relevant for domestic species with which humans play a special role. Pets such as dogs often have a special status with humans who sometimes equate them with their own species. This anthropomorphism therefore contradicts von Uexküll's notion of *Umwelt*. And for this reason, it often leads to misunderstandings that can have unfavourable consequences for the well-being of the pet.

We cannot understand the behaviour of animal species if we are not constantly aware of the adaptations that have differentiated them through evolution. Some species have conquered the terrestrial environment on or under the surface of the soil. Others have conquered the aerial environment, others the sea, and these very different environments impose constraints on the species that inhabit them. Depending on the different taxa, the way that animals have adapted through evolution to the terrestrial, aquatic or aerial environment will look the same although it may not necessarily involve the same anatomical structures. These adaptations were called by Lorenz (1974) 'evolutionary convergences' that reveal anatomical-physiological analogies. Evolutionary convergences emerged in the case of structures developed for flying in taxa as diverse as birds that have feathered wings and mammals such as bats and glider squirrels that have patagia or membranous 'wings'. All higher vertebrates, terrestrial, aquatic and aerial, are tetrapods – they have four limbs. The wing of birds evolved from the forelimb. This limb, corresponding to the human arm, forearm and hand, has feathers that increase the surface area and thus its ability to lift into the air. Mammals that fly, such as the bat, took another evolutionary course. Their lifting surface is a membrane joining the highly developed fingers of the hand, hence the name of the order of mammals that comprises bats, Chiroptera, which means 'hand-wing'.

These wide-ranging adaptations must always be considered and accounted for in experimental work, and particularly when drawing conclusions from it. Often, they pose challenges in developing experimental techniques for studies into comparative capability. For instance, someone experimenting with dogs and cats would need to consider that the forelimbs of the cat are much more powerfully developed than those of the dog. The cat uses its forelimbs, and its retractile claws, to capture prey while the dog captures and kills principally using its jaws. Even though the dog may use its forelimb, it may not do so as spontaneously as the cat. So considering these different adaptations would be vital for the experimenter in drawing clearly interpretable and unbiased conclusions.

References

Catania, K.C. (2011) The sense of touch in the star-nosed mole: from mechanoreceptors to the brain. *Philosophical Transactions of the Royal Society of London. Series B, Biological Sciences* 366, 3016–3025. DOI: 10.1098/rstb.2011.0128.

Heffner, R.S. and Heffner, H.E. (1983) Hearing in large mammals: horses (*Equus caballus*) and cattle (*Bos taurus*). *Behavioral Neuroscience* 97(2), 299–309. DOI: 10.1037/0735-7044.97.2.299.

Krubitzer, L. (1995) The organization of neocortex in mammals: are species differences really so different. *Trends in Neurosciences* 18(9), 408–417. DOI: 10.1016/0166-2236(95)93938-t.

Krubitzer, L. (2007) The magnificent compromise: cortical field evolution in mammals. *Neuron* 56(2), 201–208. DOI: 10.1016/j.neuron.2007.10.002.

Krubitzer, L. and Kaas, J. (2005) The evolution of the neocortex in mammals: how is phenotypic diversity generated? *Current Opinion in Neurobiology* 15, 444–453. DOI: 10.1016/j.conb.2005.07.003.

Krubitzer, L.A. and Seelke, A.M.H. (2012) Cortical evolution in mammals: the bane and beauty of phenotypic variability. *Proceedings of the National Academy of Sciences of the United States of America* 109 Suppl 1, 10647–10654. DOI: 10.1073/pnas.1201891109.

Krubitzer, L. and Stolzenberg, D.S. (2014) The evolutionary masquerade: genetic and epigenetic contributions to the neocortex. *Current Opinion in Neurobiology* 24(1), 157–165. DOI: 10.1016/j.conb.2013.11.010.

Lorenz, K.Z. (1974) Analogy as a source of knowledge. *Science* 185, 229–234. DOI: 10.1126/science.185.4147.229.

Menzel, E.W. and Johnson, M.K. (1976) Communication and cognitive organization in humans and other animals. *Annals of the New York Academy of Sciences* 280, 131–142. DOI: 10.1111/j.1749-6632.1976.tb25481.x.

von Uexküll, J. (1934) *Streifzüge Durch Die Umwelten von Tieren Und Menschen.* Rowohlt Verlag, Berlin, Heidelberg. DOI: 10.1007/978-3-642-98976-6.

Behaviour: How Do Animals Respond to What They Know?

3

3.1 Social Structures

Aristotle distinguished between species that led solitary lives and those that lived in groups and, among these groups, those that remained stable over several generations. He thus differentiated between solitary species, gregarious species and, within gregarious species, social species. This concept implied different degrees of tolerance between individual adults within these different species. So, in solitary species, adults are intolerant of each other except, of course, when males and females meet for copulation. In gregarious species, adult individuals are attracted to other individuals of the same species and often form large groups. However, this attraction may only be opportunistic. For example, Rabaud (1937, p. 74) described as 'crowds', groups that were only a 'collection of animals with no direct connection to each other'. He defined crowds as fortuitous and temporary, and used as an example terrestrial cave bats. Individuals of these species gather in caves but they do not create links with each other.

Tolerance between adult individuals is variously but regularly seen when animals are feeding or drinking or combining for seasonal events like migrations or mating. As Aristotle pointed out, this 'inter-attraction' is a feature of both gregarious and social species. However, in social species, unlike gregarious species, grouping of adult individuals, often females, is long-lasting. They show strong spatial cohesion, so that individuals remain much closer within their own group than with those of other groups. Within these close groups, sometimes termed 'reproductive units', they mate and take care of their young. As a result, Mason (1976) argued that, for social species, the young are both the products and the producers of social animals.

The social structure of a group is based on its composition of adult males and females. This structure was classified by Fedigan (1982), using primates as an example. She termed groups consisting of single adult males and single adult females as UM/UF (uni-male, uni-female), in other words, monogamy. And because the grouping persists for life it implies that the young of both sexes leave their parents to form their own groups when they reach sexual maturity. Monogamy is found in gibbons, a species of Asian arboreal primates, and in marmosets and tamarins, primate species from South America. We also see it in wolves, dogs and many species of birds. Sometimes, when these groups evolve in environments where resources may vary with the seasons, this monogamy may not be complete. For example, in tamarins and wolves, the young often do not disperse but, instead, become part of an 'extended family'. None the less, the core of these groups is the founding couple, and the reproductive capacity of the offspring who have not dispersed is inhibited both physiologically and behaviourally.

By contrast to monogamy, other groups may consist of a single adult male (UM) and several adult females (MF), a form of 'polygyny' (UM/MF). In these groups, the male offspring that are not involved in reproduction often create independent groups of 'single' males. This, once again, highlights the importance of intraspecific inter-attraction in such species. Polygyny is common in many wild species and in almost all domestic animals. Examples include monkeys, such as African arboreal cercopithecus and Asian colobus, horses, some zebra species, cattle and galliform birds such as chickens, turkeys and quail.

A third classification, endogamy, is seen in permanent groups of several adult males (MM) and females (MF) and their immature offspring. A feature of these groups is that adult males usually emigrate and immigrate from the group. Unlike in the gregarious species groups, most adult individuals remain within these MM/MF groups over several generations.

3.2 Social Behaviour

Social organization within a group refers to the interactions between individuals within the group. This day-to-day behaviour varies widely. Interactions can either lead to rapprochement by individuals to maintain proximity, or to protagonists seeking to keep well apart. When experimenters see repeated interactions between pairs of individuals, they can infer the nature of the relationship that binds them. This relationship can be either 'positive' or friendly, which maintains the cohesion of the group, or 'negative', which potentially may weaken the cohesion of the group (Deputte, 1983).

A manifestation of the negative component is the concept of 'dominance', a term that seems to be universally but loosely described as the relationship. But it is only part of the story. Maintenance of the group is ensured by the positive component or at least by relationships where the positive component takes precedence over the negative. Similarly, the negative component explains how conflicts are resolved within a pair. Conflicts are resolved by the flight of one of the protagonists or by clear signs of subordinate behaviour that signals to the other that it should inhibit its aggression. In many cases, this inhibitory behaviour is a resurgence of infantile behaviour because young are usually not assaulted, or are assaulted only slightly, by adults.

Our fascination with aggression is probably due to the often spectacular nature of the fights that can be the ultimate, albeit exceptional, phase of aggressive interactions. It may also be because humans see wars as inevitable in human society. That is probably why some find the aggressive behaviour of some animals 'unacceptable'. In fact, many different reactions in other living beings are often grouped under the term 'aggression'. But they only make sense in the context in which they manifest themselves.

Aggression is essentially an adaptive behaviour expressed towards any protagonist, whether or not it is from the same species. Its function is to drive the protagonist away and there may be three reasons for that: (i) to protect a resource from one of its own; (ii) to seize a resource from one of its own; or (iii) to protect itself from a protagonist that it considers a threat. The resource can be food, offspring or a potential mate. For example, after calving, cows protect their offspring from predators but also from other cows (Turner and Lawrence, 2007). So they tend to defend their young against their herd mates and against aggressors of other species. Dogs and, of course, humans do the same.

Post-parturient cows are responsible for many injuries to humans when handling cattle. Thus, in a herd of Limousine and Salers cattle, in the few days following calving a third of the cows attacked a human who approached their calves (Le Neindre *et al.*, 1999). This behaviour became less frequent thereafter. Cows also often behave aggressively towards young other than their own that try to access their udders. This aggressiveness can sometimes be so extreme as to kill the offending young. This response at the individual level contrasts with the collective defence of young under other circumstances. A barking dog in a herd of suckler cows induces many females to gather to confront the dog while the young remain protected behind them.

All males may exhibit aggressive behaviour towards other males, in the context of seeking and protecting sexual resources. As illustrated in Figs 3.1 and 3.2 aggressive behaviour may take a characteristic form. First, a fixed gaze on the protagonist, and then a rigid posture directed clearly at the protagonist. If the protagonist moves away in response to these threats the interaction ends. But if it does not, the aggressor exhibits increasingly menacing behaviour to attempt to intimidate the opponent. This gradation in the intensity of aggressive behaviour is a direct response to the reaction of the protagonist. So aggressive behaviour is fundamentally a two-way interaction. It is only aimed at distancing one of the protagonists as economically as possible and is adjusted by

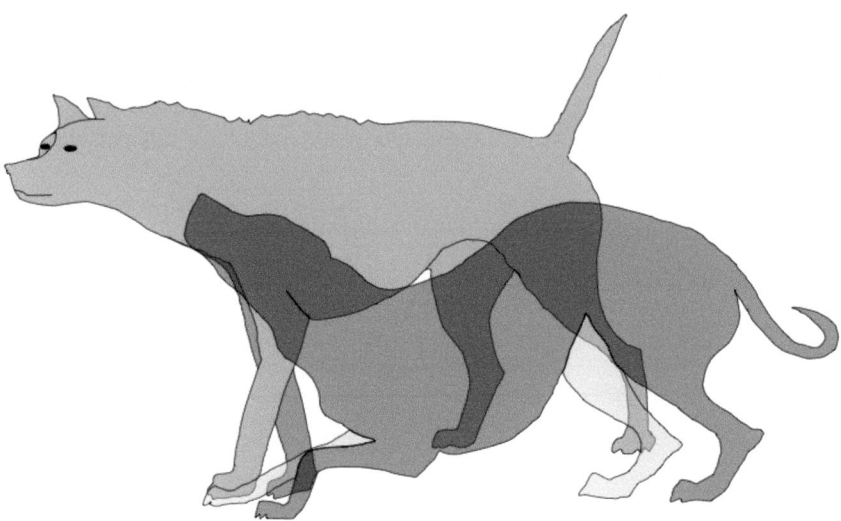

Fig. 3.1. Postures observed during agonistic interactions. These interactions include both aggressive and avoidance behaviours. Postures associated with aggressive behaviours and those associated with avoidance, flight or submission are clearly contrasted. An aggressive subject – silhouette in light grey – has its body tensely forward featuring a readiness to attack. In contrast to this posture, an aggressed individual – silhouette in dark grey – lowers its body and displays a backward movement. This indicates a motivation to flee and/or to make itself as small as possible. After Darwin, 1872 and Deputte, 2010.

the aggressor according to the response of the other. The outcome of an interaction is fundamentally asymmetrical: one assaults and the other flees. However, this asymmetry may appear only after an escalation of identical behaviour on the part of the two protagonists, whose motivations, for example to keep or to steal the same resource, are then identical – in other words, symmetrical.

Regardless of the species, some individuals express aggressive behaviour much faster and more frequently than others. These individuals are aggressive in temperament, but many others are only aggressive when necessary, such as when mothers protect their offspring. So aggression is not the same as aggressiveness. Aggression can be seen as a set of specific behavioural actions to chase away a protagonist. Aggressiveness is the propensity for an individual to exhibit aggressive behaviour (Deputte, 2007). An individual with an aggressive temperament will begin to show aggressive behaviour as soon as it senses danger, even if that danger is of low intensity. In contrast, a mildly aggressive individual will only begin to exhibit low-intensity aggressive behaviours when faced with dangers or high-intensity stimuli. Let us examine this more closely.

Fig. 3.2. Aggression. The 'bluff' is an important component of aggressive behaviours. The subject that perceives a very high and imminent danger makes itself look much bigger. Its opponent is then presented with a foe much bigger and more dangerous than initially expected. In this figure the cat displays an arched posture and looks bigger thanks to a general piloerection. At the same time, the cat moves its ears backwards and displays its readiness to attack by lifting its left forelimb. Adapted from Darwin, 1872 by B. Deputte.

3.3   Is Predation the Opposite of Aggression?

There is another form of attacking behaviour – predation. But, while aggression is a reactive and relational behaviour, predation is a part of ingestive behaviour. It is therefore a cyclical, rhythmic behaviour, the trigger of which is an internal imbalance due to hunger, but that hunger disappears after eating. Aggression and predation should not be confused because they are part of two different motivations and involve diametrically opposed behaviour. In aggression, the first manifestation is, after a fixed gaze, a movement of intention to attack, a forward extension of the body and, possibly, an increase in the size or volume of the individual, by piloerection for example. Piloerection makes the aggressor appear bigger than it is, like a cat ready to attack while cornered by a dog (Darwin, 1872; Fig. 3.2).

As we saw earlier, the function of aggression is to distance protagonists without endangering their physical integrity. In contrast, a predator, such as the cat, after staring at its prey, approaches it stealthily, its belly close to the ground. If the approach is successful, it will leap onto the prey before eating it. These two opposite forms of behaviour are controlled by fundamentally different brain structures, translating to opposite behavioural characteristics – relational on the one hand, consumption on the other (Gregg and Siegel, 2001; Deputte, 2007).

Different types of social organizations are described in the literature. Some species show a 'despotic' social organization where dominance/subordination relationships dominate the social interaction and, as a result, aggressive interactions are frequent, for example in the rhesus macaque (Thierry, 2015). But in this species, de Waal (de Waal and Yoshihara, 1983) highlighted reconciliation behaviour that still maintains cohesion of a group despite the frequency and intensity of conflicts within it. In the rhesus macaque, after an aggressive interaction, the aggressor reverts to 'friendly' behaviour.

A further partitioning of social behaviour in primates into cooperation and competition was suggested by some authors (e.g. Bernstein and Williams, 1986). A similar nuance in the social behaviour of insects and birds was proposed by Wilson (1975) in his book *Sociobiology*. He suggested that social behaviour would be based on altruism and helping other individuals that are genetically similar. He called this behaviour kin selection. All these examples illustrate that the commonly held belief that social behaviour is merely a matter of conflicts or their resolution is far too simple. Wilson's *Sociobiology* clearly illustrated that sociality is a trait that developed during evolution in a wide variety of species, both invertebrate and vertebrate. This means that sociality is only a characteristic of groups composed of individuals of the same species, with natural selection favouring offspring that possess this trait.

3.4 The Development of Social Behaviour

Most domestic animals are social species. Farmers and animal breeders have carefully maintained this characteristic – cows are kept in herds, as are horses, and groups of poultry in flocks. In some other cases, humans have hijacked the social capacities of a species to establish a relationship between humans and individual animals of that species that, although not social, is very similar. This is what we see in the human–dog relationship, which is a unique, interspecific relationship.

The domestic cat is another exception. Like most felines, it is a solitary species where adult individuals have a low tolerance or even a total intolerance towards others of their species; however, cats vary greatly in interspecific tolerance depending on the abundance and permanence of food

(Natoli, 1985). From little or no tolerance when food is widely dispersed, they can accept the presence of others when food is abundant and concentrated (Say and Pontier, 2004). This can be seen in all places where humans supply them, directly or indirectly, with food. So the domestic cat is not really a solitary species because it only has an optional tolerance.

Like dogs and many other mammals, cats have litters that are nidicolous, meaning that they must remain sheltered in nests or dens. The young, termed altricial, are born deaf, blind and immobile and they cannot thermoregulate. The behavioural development of the young is mainly through interactions among littermates. But the behaviour seen in the burgeoning playful interactions between kittens during their first 6 weeks of life gradually turns to other playthings including live or dead prey that the mother may present to them. By the time kittens are weaned at about 50 days, they have become predators. They gradually disperse from other members of the litter and eventually become solitary. However, males may not disperse but continue to live in small groups of related individuals. In dogs, on the other hand, playful interactions between members of the same litter or between unrelated young continue until much later in life and they establish social relations.

Horses, sheep, cattle and goats have developed different strategies from those of cats, dogs, pigs and rabbits during evolution. Their litters are generally small and from birth they can perceive and interact with their new environment, and don't require a nest like cats, dogs, pigs and rabbits. They are capable of locomotion almost immediately after parturition and are referred to as being nidifugous, also called 'precocial'. They only communicate with their mothers and remain close to them to be protected and fed. This means that they must rapidly learn to identify their mother among all the mothers in the group. So, in total contrast to the strategy of nesting species, the concept of attachment evolved because the young depended on a single individual to provide contact, protection and food. The young should remain in close contact with this individual, and shows signs of distress when it loses it.

In humans, the type of attachment is intermediate between that of nidicolous and nidifugous species. The human child can interact from birth with its environment, especially with its mother, but can't move independently for many months. In non-human primates, especially monkeys and apes, the young cling to their mother's abdomen for the first weeks of their lives but they develop locomotor skills much faster than humans.

Socialization involves two fundamental processes. First, young animals develop specific behaviour through interactions with other young and with adults of the same species, both male and female, within the social group. Second, they construct a network of relationships between themselves and all the members of the group (Deputte, 2000). These relationships take different forms depending on the nature of the interactions. During the course of socialization, the young's special relationship with its mother gradually evolves to look eventually like the 'friendly' relationships that we see between young and adult females.

The term 'attachment' is currently widely used. It refers to any form of close, and more or less exclusive, relationship between two individuals regardless of species. Its use in this sense should not be confused with the original concept of 'attachment' described by Lorenz (1941) as an adaptive trait of nidifugous species. This attachment in nidifugous species is the mammalian equivalent of a 'footprint' (Lorenz, 1941). None the less, these two definitions, whose origins are different, have the same purpose. They describe an exclusive link. This link can be established between the young of a species and any individual of any other species, or even a foreign object or 'surrogate mother'. Harlow (1965) and Harlow and Harlow (1965) have described the attachment of young monkeys to wire covered, or not, with soft cloth. However, Harlow also observed that when the young imprinted with an individual of another species or with a foreign object as a surrogate mother, its behavioural development was impaired to an extent that later compromised its safety and protection. So, to develop harmoniously, the new-born must interact with individuals of its own species. In mammals, the two crucial stages of this development are birth and weaning. In the first hours after birth, the young must learn to recognize its mother, so that it can keep in close and exclusive contact. Then, after spending several months maintaining close contact with its mother and being protected, fed and comforted by her, the contact is ruptured when it is weaned. This is a powerful change because the young's psycho-physiological balance is disrupted.

However, the short periods around birth and weaning in domestic cattle, sheep and horses present opportunities for humans to establish a lasting friendly relationship. If humans are present at the birth of young for the first hours after birth, a long-standing familiarity can be established. In some cases, humans can even become surrogate mothers. The same goes for weaning. In cattle, friendly contact at weaning and soon after leads the weanling to accept or even be preferentially attracted to the human. This has been shown even if the calf has been exclusively with the mother for all of its previous life. But it is no longer possible if contact is delayed until a week after weaning (Boivin *et al.*, 1992).

Birth and weaning are periods that have been described as 'critical' then as 'sensitive' and 'privileged'. All of these descriptions are far too restrictive and do not correspond to biological reality. Moreover, the term 'period' does not imply that there are specific time-markers determining for a species, or even for a set of species, a start date and an end date. Development is fundamentally a continuous process in which different organs develop and specialize at different dates and rhythms, and for varying durations. All these processes show a high degree of individuality, and each individual develops at its own pace.

In vertebrates, an individual develops a collaborative behaviour in the specific environment in which it is growing. It is learned, so both it and its timing are likely to differ from one animal to another. Collaborative behaviour is essential to sustain social groups and to respond more effectively to challenges from the environment. Such interactions and collaboration between individuals are the basic element of social organization.

Living in groups and constantly interacting inevitably involves solving conflicts. Agonistic behaviour, which includes behaviours such as aggression, flight, avoidance and submission, make it possible to resolve these conflicts without jeopardizing the cohesion of the group and ultimately its survival. They involve the appropriation or protection of nutritional, sexual or social resources that are a normal part of the animal's life. But such conflict often shocks human observers who do not understand the natural role of most attacks and conflicts, which seem to them violent and immoral. However, it is unreal to seek to place human moral values on animals, when we should be looking to understand the advantage that behaviour gives to animals to allow them to live well and ensure the perpetuation of their species.

References

Bernstein, I.S. and Williams, L.E. (1986) The study of social organization. In: Mitchell, G. and Erwin, J. (eds) *Comparative Primate Biology. Volume 2A: Behaviour, Conservation and Ecology*. Alan R. Liss, New York, pp. 195–213.

Boivin, X., Neindre, P.L. and Chupin, J.M. (1992) Establishment of cattle–human relationships. *Applied Animal Behaviour Science* 32(4), 325–335. DOI: 10.1016/S0168-1591(05)80025-5.

Darwin, C. (1872) *The Expression of the Emotions in Man and Animals*. John Murray, London, 372 pp.

Deputte, B.L. (1983) Ontogenic development of dyadic social relationships: assessing individual roles. *American Journal of Primatology* 4(4), 309–318. DOI: 10.1002/ajp.1350040403.

Deputte B.L. (2000) Primate socialization revisited: theoretical and practical issues in social ontogeny. Advances in the Study of Behavior 29, 99–157.

Deputte, B.L. (2007) Comportements d'agression chez les vertébrés supérieurs, notamment chez le chien domestique (*Canis familiaris*). *Bulletin de l'Académie Vétérinaire de France* 60, 349–358.

Deputte, B.L. (2010) Communication, perception et expression du chien. In: Bedossa, T. and Deputte, B.L. (eds) *Comportement et Éducation Du Chien*. Educagri, Dijon, France, pp. 355–424.

de Waal, F.B.M. and Yoshihara, D. (1983) Reconciliation and redirected affection in rhesus monkeys. *Behaviour* 85(3–4), 224–241. DOI: 10.1163/156853983X00237.

Fedigan, L.M. (1982) *Primate Paradigms: Sex Roles and Social Bonds*. Eden Press, Montréal, Canada, 386 pp.

Gregg, T.R. and Siegel, A. (2001) Brain structures and neurotransmitters regulating aggression in cats: implications for human aggression. *Progress in*

Neuro-Psychopharmacology & Biological Psychiatry 25(1), 91–140. DOI: 10.1016/ s0278-5846(00)00150-0.

Harlow, H.F. (1965) Total social isolation: effects on macaque monkey behavior. *Science* 148(3670), 666. DOI: 10.1126/science.148.3670.666-a.

Harlow, H.F. and Harlow, M.K. (1965) The affectional systems. In: Schrier, A.M., Harlow, H.F. and Stollnitz, F. (eds) *Behaviour of Nonhuman Primates: Modern Research Trends.* Academic Press, New York, pp. 287–334.

Le Neindre, P., Trillat, G., Garel, J.P., Verdier, M. and Grignard, L. (1999) Aggressiveness after calving and docility of suckling cows. In: Boe, K.E., Bakken, M. and Braastad, B.O. (eds) *Proceedings of the 33rd International Congress of the International Society for Applied Ethology (ISAE),* 17–21 August 1999, Lillehammer, Norway. ISAE, Agricultural University of Norway, Lillehammer, Norway, 181 pp.

Lorenz, K. (1941) Vergleichende bewegungsstudien an anatinen. *Journal of Ornithology* 89, 194–293.

Mason, W.A. (1976) Primate social behaviour: pattern and processes. In: Masterton, R.B., Bitterman, M.E., Campbell, C.B.G. and Hotton, N. (eds) *Evolution of Brain and Behaviour in Vertebrates.* Lawrence Erlbaum Associates, Hillsdale, New Jersey, pp. 425–455.

Natoli, E. (1985) Spacing pattern in a colony of urban stray cats (*Felis catus* L.) in the historic centre of Rome. *Applied Animal Behaviour Science* 14, 289–304. DOI: 10.1016/0168-1591(85)90009-7.

Rabaud, E. (1937) *Phénomène Social et Sociétés Animales.* Félix Alcan, Paris, 325 pp.

Say, L. and Pontier, D. (2004) Spacing pattern in a social group of stray cats: effects on male reproductive success. *Animal Behaviour* 68(1), 175–180. DOI: 10.1016/j.anbehav.2003.11.008.

Thierry, B. (2015) Le rôle des contraintes dans l'évolution des sociétés de macaques. *Bulletin de l'Académie Vétérinaire de France* 168(3), 203–208.

Turner, S.P. and Lawrence, A.B. (2007) Relationship between maternal defensive aggression, fear of handling and other maternal care traits in beef cows. *Livestock Science* 106, 182–188. DOI: 10.1016/j.livsci.2006.08.002.

Wilson, E.O. (1975) *Sociobiology: The New Synthesis,* 1st edn. Belknap Press of Harvard University Press, Cambridge, Massachusetts, 697 pp.

How Do We Measure the Behavioural and Cognitive Ability of Animals?

4

4.1 Cognition in Animals

Cognition is the mental action or process of acquiring knowledge and understanding through thought, experience and the senses. It is clearly a complex concept, as many authors have acknowledged. 'Cognition brings together knowledge, approximations and errors, at the same time as the mechanisms or processes by which all these are elaborated' (Le Ny, 2002, pp. 70–71). Or, 'Cognition involves the intervention, in a given individual, of learning processes and information processing' (Vauclair, 1992, pp. 34–35). Vauclair also adds: 'In the same vein as cognition, we find the concept of representation'. He concludes: 'We therefore speak of cognition when an individual finds answers that solve a problem it has in the environment that surrounds it.'

Cognitive ability allows individuals first, to learn about their diverse environment and then, how to live best within it. The environment contains both motionless objects and animated objects. In social species, many of these animated objects are members of the individual's own species. However, both types of cognition, non-social and social, are in fact similar because they differ only in the properties of the objects, social or not, and in the relationship that the subject has with these objects, especially in the way it reacts to them.

Understanding the objects in an environment requires the capacity to memorize. However, there is a limit to the potential number of objects that any individual can identify and respond to. This means that discrimination has to be an important part of cognitive ability and individuals must distinguish between objects despite there being sometimes very close and profound similarities. In addition, discrimination may sometimes be demonstrated at a very early age. Human babies as young as 24 h old are able to discriminate by touch between a cube or a cylinder (Streri *et al.*, 2000). Streri *et al.* repeatedly put a cube in a baby's hand without the baby being able to see it. Each time, the baby manipulated it for a time and then

© CAB International 2022. *Interacting with Animals: Understanding their Behaviour and Welfare* (Pierre Le Neindre and Bertrand L. Deputte) DOI: 10.1079/9781800622418.0004

released it. The duration of successive manipulations decreased before the cube was released. Then they placed a cylinder in the baby's hand and the baby handled it for as long as it had initially handled the cube. So the baby perceived that the cube and the cylinder were different, prompting a paradigm that Streri *et al.* called 'habituation and dishabituation'.

Another component of cognitive ability, described by J. Vauclair (Aix, France, 2002, personal communication) as 'cognitive economy', is categorization. This is where animals group many objects that have common characteristics into a set, that is either 'perceptual', such as objects of the same colour, or 'conceptual', such as objects that are edible. So an individual does not need to remember all the objects in its environment, but just a limited number of relevant classes of these objects. Categorization has been demonstrated in only a few animal species due to the experimental difficulty of testing it; to do so requires protocols used in experimental comparative psychology. Those are not too difficult to handle in humans but very difficult in animals. Broadly, this means teaching animals to respond in a measurable way, for example pulling on a rope, moving towards an image or screen, or pressing a button to one class of objects and responding in another way or not responding at all to another class of objects. These protocols relate to a simultaneous discrimination between two objects of two different classes. They require that the answers are not ambiguous, so subjects are given a reward, usually food, when they provide a correct response to the criteria set by the experimenter. These answers are therefore learned. So drawing conclusions from experiments into whether or not animals have the capacity to 'categorize' is only possible after a great diversity of stimuli has been presented to them. As a result, categorization has been demonstrated in only a few species although it is likely that many species of higher vertebrates and invertebrates are capable of it.

Animals can become aware of objects using a single sensory mode or several modes and this varies with species. However, most of the published experimental information emphasizes vision. Only a few focus on sound or smell. Nevertheless, an animal is potentially able to recognize objects and have a mental representation of them. The object itself can be recognized through a single sense, or a combination of several of these senses which shows that an individual's world, its *Umwelt*, can be very complex – and we know frustratingly little about it. None the less, we must assume that many animal species, domestic and wild, benefit from a rich knowledge of the objects in their world. By the way that animals interact with these objects and because they are in constant contact with some objects throughout a year or a lifetime, animals may also be aware of relationships between objects themselves (Doré and Mercier, 1999).

What applies to inanimate objects in the environment also applies to social species that form permanent groups. Differentiated relationships within a social group depend on individuals recognizing others of their group. This seems a reasonable hypothesis, but surprisingly, it has only been validated in very few species. This is because when experimenters use vision to study the nature of the stimuli presented, they must be conscious that the individual animal has to recognize another of its species in the form of an image. It should respond consistently when it sees it from different angles – face on, in profile, three-quarters from behind or in front, and differentiate any of these images from those of other congeners. This has been shown by Autier-Dérian *et al.* (2013) in dogs and Coulon *et al.* (2009) in cattle.

Long-tailed macaques are even more skilful: two subjects could relate just a few body parts to an individual, including matching its thigh to the corresponding face (Dasser, 1987). There are similar reports on visual recognition in wasps (Tibbets, 2002) and acoustic recognition in penguins (Lengagne *et al.*, 1999). But there are many more unsubstantiated reports on the ability of animals to recognize individuals than reports that have been unequivocally demonstrated.

In social species, these abilities are part of the dynamics of their social relationship. Social relationships are based on alliances between individuals, particularly within age groups and the sexes. Social cognition is thus an integral component of all decision making during interactions, especially when it must be rapid. The animal must learn the motivation of a partner, adjust its own response to it, and communicate its own motivation. The complexity of this social dynamic in chimpanzees is described by de Waal (1984) in his book *Chimpanzee Politics: Power and Sex among Apes.*

There are many examples in the literature on non-human primates that show how some individuals can manoeuvre to modify the behaviour of others in social life. Byrne and Whiten (1988) grouped them under the explicit term **Machiavellian intelligence**. They described the case of a young male baboon watching an adult eating particularly appetizing yams to which he was strongly attracted. He then began to make loud and high-pitched screams, the cries that individuals emit when they are attacked to thwart an aggressor and attract the attention of congeners. In this case, the adult with the yam was surprised and left it, since he had learned that these cries emitted by a young, make the mother rush to protect it. The young baboon then grabbed the yam. Events like this are usually unpublished and carefully stored in the filing cabinets of primatologists because they are not quantifiable and difficult to reproduce experimentally. Nevertheless, they illustrate the complexity of cognitive abilities based on individual learning during interactions or after observing interactions between partners.

Relatively recently, a new capacity has been highlighted among social animals called the theory of mind (Premack and Woodruff, 1978). It is the ability of individuals to attribute knowledge or other competencies to their partners, or in other words to feel the possible feelings of others. Authors in the past have suggested that this is an attribute that is unique to humans, but we now have experimental evidence that it exists in chimpanzees and a few other species of higher vertebrates. It is clearly a form of high-level consciousness.

Research into cognition of both social and non-social animals has grown considerably in the past 30 years. That is particularly so in domestic and companion species like the dog (Miklosi, 2015). This is partly because the dog can develop a strong relationship with humans and heeds their instructions. The cat, on the other hand, is uncooperative in an experimental context. Does this mean that the cat has fewer developed cognitive abilities than the dog? In fact, as yet no one has found a way to deny or confirm this. Some argue that the cognitive abilities of cats are highly developed but so far there is little or no scientific supporting evidence. The accumulation of anecdotes about the performance of the cat is no substitute for scientific data. In fact, it is going to take a lot of imagination for researchers to induce cats to cooperate in experimental protocols, without altering their underlying behaviour.

Behaviour expressed by animals is a combination of the evolutionary history of their species, or phylogenesis, and their personal history, or ontogenesis. It is under control of both the nervous and hormonal systems. These systems, in turn, vary with the animal's physiological stage and age. Thus, during periods of sexual activity the behaviour of both males and females is specific for each species. Males, apart from being attracted to females, become aggressive towards other males to a degree that is often proportional to the increase in their concentration of male sex hormones. This aggressiveness leads to increased fighting and can extend to other species which makes them more dangerous to humans at these times. Contrary to what some authors claim, females will often seek a male, whether they accept copulation or not. Fabre-Nys and Venier (1987) give an example of that form of proceptivity.

At parturition and afterwards, female sheep change their behavioural patterns substantially (Poindron and Le Neindre, 1980). At all times apart from lambing, ewes are indifferent and occasionally aggressive towards young lambs. But at lambing, they are attracted to new-born lambs, lick them and facilitate their access to the udder (Levy *et al.*, 1983), so much so that they often become attracted to new-born lambs of other ewes and not just their own. But this indiscriminate maternal behaviour is quickly replaced by a selectivity in which the ewe only allows her own or an early-adopted young to suckle (Poindron *et al.*, 1984). There is thus a short sensitive period when the ewe will adopt one or more lambs and accept them at the udder. This vital change to the chance of the lambs' surviving is dependent on, or mediated by, neuroendocrine changes especially involving oestradiol (Poindron *et al.*, 1986). The massive and very rapid change from non-maternal to maternal behaviour is triggered by vaginal stimulation during the passage of the young through the genital tract. In fact, this maternal behaviour can be initiated artificially in ewes simply by stimulating the genital tract of non-parous, non-lactating ewes (Keverne *et al.*, 1983). Outdated concepts such as 'maternal fibre' are no longer relevant in sheep. Some females develop strong bonds with some young, but these females may not necessarily be the biological mothers of the lambs they rear (Poindron and Le Neindre, 1980).

Hormonal and nervous stimulation of cognition has also been demonstrated in other species, but the determinants can be very different. For example, rats, both male and female, when first shown new-born pups, may even eat some of them, but they progressively and rapidly adopt the characteristic behaviour of post-parturient females (Rosenblatt *et al.*, 1979). They make a nest, gather the pups together in this nest and adopt a suckling posture even if they are not lactating – or are even male!

In pigs, distinctive sexual interactions between partners were demonstrated by Signoret and his team (Signoret, 1970a, b, 1975). They described the intricate steps that lead to the sow being fertilized by the boar. The presence of the sow induces the boar to secrete saliva containing substances that originate from male sex hormones. These substances provoke the sow to remain immobile for several minutes during mating by the boar which, in pigs, is necessary to enable successful copulation. Moreover, the presence of a boar advances the age to puberty of young sows, a mechanism that is called the 'male effect' (Prunier, 1989). The term 'male effect' was first used long before 1989, in particular for ewes (Thimonier *et al.*, 2000). These complex sequences contrast with the simplistic assertion of Nicolas (2020, p. 152) that 'The sad reality of the animal world is that most females are raped, and most males are rapists'!

The behaviour we observe in animals results from the activities of the central and peripheral nervous systems, stimulated by sensory receptors, and modulated by the hormonal system. Together, they make up the social life of individuals. Nervous tissue consists mostly of neurons, or cells that transmit and process electrical signals. They are surrounded and supported by glial cells that monitor their chemical and electrical environment. Glial cells provide neurons with nutrients and eliminate waste products. This complex association is beyond the scope of this book, but we still need to examine some aspects of these components to help understand the way that animals behave.

In vertebrates, the nerve structures include, of course, the brain and spinal cord which are under central control. However, the autonomic, peripheral nervous system also has neurons and is not directly subject to central control. It controls the movements of the digestive system but also its secretions and vascularization. Some nerve relays such as sight, olfaction, hearing, touch, and pain receptors transmit sensory information from the periphery to the central system. This information usually undergoes initial treatment, so that for example, when an animal perceives an odour, it has already been processed to eliminate olfactory signals that it deems irrelevant and only delivers those that might alert the animal and allow it to decide on acting. Pretreatment of signals like this can result in very rapid, so-called **reflex actions** that do not involve the central nervous system at all, but just the brain stem, including the spinal cord. Later, the central brain system may become aware of the action rather than the signal. We see this too with tactile and balance receptors that can alert the whole body of the animal to changes in the surface on which the body is moving. An extreme and complex example of this is **air-righting** of cats, in which their reflex is to orientate themselves instantly during a fall to ensure that they land on their feet. The nervous system manages how the different parts of the individual relate to each other and how the individual relates to the outside world.

The brain of vertebrates has structures that are dedicated to processing information and memorizing it. Two structures are of particular interest here. The first is the cortex which takes on great importance in humans and some great apes but exists in all vertebrates. The second is the limbic system, which is more specifically dedicated to the creation of emotions. Other structures have a role in the management of endocrine and metabolic functions.

This overview of the nerve system and cognition leads to the concept of consciousness, also sometimes called awareness. In some situations, the brain is not able to analyse what is happening around the animal. In other words, the animal is unconscious. But there are levels of unconsciousness in both humans and other animals (Laureys *et al.*, 2004). In abattoirs, a state of unconsciousness, or stunning, is induced in animals before they are killed. Several stunning procedures are used, but all have the purpose of ensuring that animals suffer no pain before they die.

4.2 Consciousness in Animals

Consciousness in animals has been debated in the scientific and philosophical literature for a long time. It is linked to the immediate, or proximal, causes of animal behaviour rather than the long-term or distal causes. Behavioural ecology is particularly concerned with the distinction between the so-called distal (or ultimate) causes and the so-called proximal causes of animal behaviour (Tinbergen, 1963). Ultimate causes refer to the mechanisms that, during the evolutionary history of a given species, have led to the appearance of a particular behaviour within the behavioural repertoire of the species in question. Proximal causes refer to the reasons why, in a particular context, a particular animal will behave in a particular way.

One school of thought considered, and sometimes even today still considers, that the behaviour of animals results from reflex responses to sets of stimuli and that higher nerve centres play little or no role. If this were the case, then a complex ability such as consciousness would not be relevant. But it has long been known that reflex stimuli-responses contribute strongly to human behaviour, and probably to that of animals. It is also known that these reflex acts are not necessarily produced by the brain but by other nerve structures like the spinal cord. None the less, other people say that nerve centres are strongly involved in the making of complex responses to the environment (Damasio, 1999).

If behaviour in animals were only the result of stimuli-responses, then Descartes's (1637) description of an animal as a machine would be adequate. However, later authors believe that animals have complex mental states that are personal to the individual and are not directly accessible to other individuals (Le Neindre *et al.*, 2017) except in the case of humans because they can use language to communicate. This is not a tenable argument because we know that, in animals, as well as humans, a lot of information is transmitted by means other than language. It would also imply that humans could not access the mental states of infants and the mentally impaired. In 2012 scientists working in this area (Low *et al.*, 2012) postulated that animals have mental states and some form of consciousness. The French National Institute for Agricultural Research (INRA) assembled a group of senior scientists and philosophers in 2017 to collate the available information on animal consciousness based on the idea that mental states are the product of nerve activity and do not come from theological or other sources. We summarize their findings here.

Apart from reflex responses, the brain processes information based on what it has experienced in its past, in other words what it has learned from previous activities and whether it had a positive or negative emotion as a result. Understanding this process, apart from its academic value, is thought to be the key to our deciding the best way to improve the quality of life of animals and to prevent their reacting unfavourably towards themselves, their congeners and humans. Research to uncover this information has not been easy. People who care for animals often consider them as objects and readily associate the animals' responses with what their own might be. This seemingly intuitive understanding may have the advantage of efficiency but can lead to false interpretations and inappropriate decisions about management. It is anthropomorphism, or assuming that our own way of reacting would be the same as that of animals by 'putting ourselves in their place'. It can also lead to anthropocentrism, in which we analyse a situation according to our own wishes and desires as if we were in the same situation as the animal. So these approaches have their limitations. We are uncertain whether what the human experimenter or carer tells us is the thoughts and desires of the animals or of the human. In the face of this difficulty, 19th-century researchers simply considered the brains of animals as a black box from which they were only concerned with outcomes and did not attempt to interpret the underlying causes of the responses they observed. These outcomes could have been a movement that was visible but could also have been changes in the animals' basic physiology. There are many such changes, but those that attracted the most attention from researchers were stress reactions and the hormones associated with them (Dantzer and Mormède, 1979).

Some of these stress hormones act on the entire physiology of the animal. Thus, the concentration in the blood of a hormone like cortisol can change the balance of many metabolic processes such as metabolism and secretion in the bloodstream of sugars, lipids and proteins as well as immune mechanisms. Tissues can also react to cortisol which constricts certain blood vessels to direct blood to flow to specific muscles or viscera. The creation and secretion of stress hormones is dependent on a complex set of multiple brain structures, including the pituitary gland and the adreno-cortical glands. Those complex changes are, however, only consequences of the way that animals perceive their environment.

Therefore, much of the research in ethology and comparative psychology assumes that animals have mental states and mental representation. It is surprising that, even today, some people question this by relegating animals to the category of machines. Such a stance, which was understandable in the 17th century given the state of knowledge at that time, seems untenable in the 21st century.

None the less, information on mental states is indirect because we still haven't found a way to measure quantitatively what happens in the brain. It should be noted though, that most of what we know about brain function in humans would be invalid if we only relied on quantifiable elements. Even though humans can speak, their preferences, pleasures and fears can only be evaluated indirectly. There are no quantifiable standards for these feelings, but we believe in them because we can feel them ourselves. To study the mental states of animals, scientific methods that did not rely on language were used. They all assumed concepts tested in the field from assumptions defined *a priori*. The mere accumulation of anecdotes is not enough to provide credible evidence of hypotheses concerning mental status. That is why, over time, an increasingly demanding methodology has had to be developed.

An animal's ability to analyse its personal experience is defined as its consciousness or awareness – the subjective experience that the animal has of its environment and its own body. This definition does not infer any moral value related to its behaviour – attributing moral values is the realm of ethicists not scientists. Referring to an animal as having a 'bad conscience' or a 'good conscience' is not the purpose of scientific literature. Reflecting on their own actions in terms of morality has long been the domain of humans. That way of thinking is not the one we want to pursue.

The boundary between humans and animals has been questioned for centuries to try to understand it, but also for theological reasons, often with the aim of placing humans apart and more particularly, above animals. Perhaps this is why humans want to justify the control, and the unbridled exploitation that humans want to have, and often do have, over animals.

For a long time, researchers working on animals had no tools to test whether animals had skills such as language that were supposed to indicate consciousness in humans. Therefore, few people thought it worthwhile to test these skills in animals since animals have no language. There are reasons why animals do not act as humans might expect. The first is that the questions asked of animals probably do not make sense to them. Second, even if the questions make sense, the animals may be reluctant to answer them due to the context in which they are asked.

In general, methods used to explore animals' responses to meaningful situations are both behavioural and neurobiological. Researchers use peer-reviewed techniques, and, like all good scientific analysis, their results test hypotheses formulated before the experiments begin. Nowadays, it is normally accepted that conclusions cannot be drawn from fortuitous observations and anecdotes. This was not the case in the literature of the 19th century which abounds with anecdotes (Menault, 1868). Many books about the wonders of behaviour adorn library shelves. However, even though interesting, this anecdotal material is only useful for generating hypotheses, so perhaps we can still call it valuable!

Some people may consider that the difference in cognitive skills between humans and animals makes it impossible to formulate hypotheses for testing in animals based on results in humans. This idea can be challenged on two grounds. The first is that the concept is unsound, because comparing does not infer equivalence. The second is that hypotheses are the result of creative activity without being solely a matter of reason and therefore unscientific. Formulating hypotheses appeals to the imagination, whether poetic, theological or social. In fact, we believe that any hypothesis is admissible provided that it is subsequently tested against reality. Paradoxically, some people who deny that animals have mental ability successfully use results from animals to study neurobiological mechanisms in humans to propose innovative therapeutics.

As we said earlier, this reluctance to use the results obtained in humans is usually accompanied by another reluctance to use the same vocabulary as is used to describe behaviour and mental state in humans. In fact, we think it is pointless to create a vocabulary that is meaningless for humans to describe objectively the subjective feelings of animals. For example, it is difficult to replace the term **pain** with another word, other than to describe all the reactions one sees such as increased heart rate, the flight or attack reactions and stress hormones in the bloodstream, like cortisol or cortisone or adrenaline and enkephalin. On the other hand, the word 'pain' includes all these reactions, regardless of whether they involve humans or animals, but they are often the result of the brain's analysis of a given situation, not its cause. Moreover, words that suggest a level of anthropomorphism are not avoided by researchers from other disciplines. So physicists speak of the 'charm' and 'colour' of fine particles, without assuming that this has an anthropomorphic connotation. It seems to us that the only important thing is that the word one chooses can be used to give an operational definition and a qualification or, better still, a quantification of what one is trying to study.

In this context, de Waal (2016) looked for signs of consciousness in animals using clearly described methods and protocols. He came up with two ways to describe consciousness. One refers to the level and the other to the content.

4.2.1 Consciousness and unconsciousness

The level of consciousness of an individual is based upon its ability to perceive its environment and to perceive where it sits in that environment. The environment for humans may be as diverse as full wakefulness, death, coma, deep sleep, rapid eye movement (REM) sleep, dreaming, sleepwalking and anaesthesia. In animals we can recognize the same typology although, for the moment, it is less varied than in humans. Animals may be dead or unconscious. They may recognize and react to internal signals, such as hunger or heat and cold, or to external signals, such as the sight or smell of a predator, prey or social partner. They can dream (Pena-Guzman, 2022) – who hasn't seen a young puppy make sucking movements while it is sleeping? And, during dreaming certain parts of the brain whose importance for consciousness have been characterized in humans show nerve activity just as they do in humans. This is not merely a subject of academic interest, any more than it is in humans. We are all aware how delicate, and even painful, it can be when we differentiate states of unconsciousness associated with different types of comas from the state of consciousness. This distinction is also important in animals to assess their state of unconsciousness during anaesthesia. Veterinary surgeons have to know how to anaesthetize animals of many species and render them insensitive. The object is for the surgeon to be able to operate in safety, but especially to avoid significant pain for the animals.

Effectively rendering animals unconscious is also important when slaughtering animals in abattoirs. To reduce suffering, major animal welfare regulations in most countries require that animals be rendered unconscious before killing them. It is not possible for human health reasons to use anaesthetics, and at slaughter, it is not necessary to know the animals' cognitive processes, but to ensure that these processes are destroyed. Three methods are usually used depending on the species concerned:

1. Mechanical stunning or a violent blow to the head, usually with the penetration of an iron rod into the skull. This method is used in large animals such as cattle, pigs and horses. Unconsciousness is usually instantaneous but, if not, the process is repeated.
2. Electrical stunning after placing electrodes on either side of the skull. The electrical current generates an epileptic seizure that causes

unconsciousness. Often, a third electrode is placed on the back, resulting in cardiac fibrillation and heart failure. These techniques are mainly used to stun pigs and poultry but are sometimes used with calves. The results appear satisfactory for pigs, but less effective for poultry principally due to the rates of slaughter required.

3. Gas stunning or placing animals in mixtures of inert gases, such as carbon dioxide, nitrogen or argon. After a period during which they may be agitated, the animals lose consciousness. This technique is used mainly in pigs and poultry.

In contrast, some religions, for example Islam and Judaism, require that animals are not stunned at slaughter. Their religious dogma demands that, for Jews, animals must be conscious or for Muslims they must simply be alive at slaughter. Beyond the dogma, members of these religions argue that slaughtering without stunning causes less pain and anguish for animals than the techniques that include stunning. This view is obviously very controversial, so that many animal welfare societies, and veterinarians responsible for verifying the effectiveness of the process, are calling for this type of slaughter to be abolished.

4.2.2   The make-up of consciousness

Apart from the level of consciousness, we can also analyse the components that make up the concept of consciousness. For a very long time, it was accepted that only in humans was consciousness made up of various components and associated with moral competence, with the implication of good and evil. But that is irrelevant for animals. However, over the past 20 years new information has challenged the dogma that only humans have a consciousness that enables them to reflect about themselves and what they do. We will briefly analyse here some of this work in animals on issues such as emotion, metacognition, time travel and relationships between individuals.

If we look beyond the simple response of animals to stimuli, we can discern their emotions in response to these stimuli and whether an animal is negatively or positively interested in them. These emotions imply that the animal has analysed its situation in relation to expectations that do not always require complex cognitive processes (Boissy *et al.*, 2007). In sheep behavioural and physiological responses to a stimulus have a subjective component. This includes the stimulus's relevance and its implications for the individual, the possibilities of action and its relationship to normal behaviour, all of which modulate the animal's ultimate response. Animals use their past experiences to decide their responses to situations in the present. So, like humans, they can be considered emotional. Positive or negative experiences in sheep determine what they learn and memorize. This leads them, for example, to have optimistic or pessimistic attitudes, or cognitive bias, as it is termed, when confronting new situations (Destrez *et al.*, 2013, 2014). Such cognitive bias has also been demonstrated in pigs (Douglas *et al.*, 2012).

There is also clear evidence of episodic memory in animals. Episodic memory has long been considered to be unique to humans. It describes the fact that we can remember specific events in our lives. It differs from semantic memory, which relates to the knowledge of rules, such as 'I get up in the morning' or 'the sun rises in the east'. Some humans may have semantic memory without having episodic memory. Because there was no evidence of episodic memory in animals some authors concluded that animals were 'stuck in the present' (Roberts, 2002). Until recently this position was also widely shared by philosophers and scientists. It was thought that mastering the past and the future was a quality that only adult humans could convey in their memories and expectations. Doubts about this belief came from people who realized that it was not testable in animals because they did not have a language to explain their memories. So it was necessary to define methods that gave testable results without verbal language. To make hypotheses testable in animals, some researchers concluded that an event could be characterized by three components: (i) what; (ii) where; and (iii) when. This is abbreviated to WWW for *what, where, when.*

The first examples using this approach were from experiments on a North American corvid, the western scrub jay (*Aphelocoma californica*) (Clayton and Dickinson, 1998; Raby *et al.*, 2007; Raby and Clayton, 2009). Birds from this species live in an environment where resources are scarce and seasonal, and a feature of their behaviour is that they hide food. In an experimental environment, these birds prefer to consume insect larvae, which degrade rapidly, rather than seeds, which do not degrade over time. A simple but ingenious experimental protocol made it possible to test the authors' hypotheses. Animals living in cages had the opportunity to hide either insect larvae or pellets in sand-filled bins. After letting them hide food of both types in the bins the birds were removed from the cage. Then they were put back after either a few hours or after several days. After several days, the larvae were dead and therefore inedible. After the short absence, the birds preferentially searched where they had hidden larvae. But, after the longer interval, they searched only where they had hidden the grain. They had thus demonstrated their ability to respond to the WWW paradigm (Clayton and Dickinson, 1998). Although this conclusion was questioned at the time (Süddendorf and Busby, 2003), further data from observations in the wild and experiments in the laboratory confirmed this conclusion and have led to its being accepted. Similar protocols were developed in other species that also hide food, including rats (Eacott *et al.*, 2005), pigs (Kouwenberg *et al.*, 2009), rufous hummingbirds (Jelbert *et al.*, 2014), chattering magpie (Zinkivskay *et al.*, 2008) and prairie voles (Ferkin *et al.*, 2008). For this, the authors took advantage of the different behavioural characteristics of the animals with which they worked. For pigs and rats, for example, the key element was the attraction of these animals to a novelty. In voles, it was the attraction to females during periods of sexual activity. Thus, episodic memory is not related to a sense or behaviour, but to the intrinsic fundamental ability of many species of animals to memorize specific events.

Several researchers have also shown that episodic memory is important for planning for the future (Raby *et al.*, 2007; Roberts, 2007; Raby and Clayton, 2009; Crystal, 2010; Eacott and Easton, 2012). Being able to create rules from past events makes it possible to use these rules to predict the future. For example, woodpeckers and squirrels hide fruit for consumption in times of scarcity. It was argued, however, that this behaviour might only reflect an ability to hide food that they might have consumed immediately. It was therefore not possible to conclude that they were doing so to prepare for the future. But further experiments, in particular with western scrub jays showed that they could plan their future (Raby *et al.*, 2007). The western scrub jays had at their disposal two types of food, peanuts and kibble, and they normally consume a mixture of both. On a given day, each bird was placed in a cage with two compartments and was able to hide food in both compartments. In the evening, the animal was confined to only one of the compartments, and the next morning it received only one type of food. The next day it was able to hide the two foods again in both compartments and, in the evening, confined in the same compartment as the day before. The bird preferentially hid the food to which it had not had access the day before in the compartment where it had been confined. Another experiment showed that the grey-headed marten, a South American mustelid, could plan for its future (Soley and Alvarado-Díaz, 2011). These animals are fond of ripe plantains. They hide these fruits when they are too green to be edible and only recover and consume them after they ripen.

Metacognition, or the ability to 'know what one knows' is another facility that is believed to be specific to consciousness. Humans may say 'I know this' or 'I have it on the tip of my tongue'. This type of path is difficult to trace in animals – again because they can't speak. However, once again, ingenious protocols have made it possible to recognize metacognition in some species. For example, dolphins can learn to press buttons and when Smith *et al.* (1995) offered sounds of two different frequencies to bottle-nosed dolphins they were rewarded if they pressed the button corresponding to the 'right' sound and nothing if they chose the other. At the beginning of the tests, the two sounds were very different, but afterwards they were deliberately made increasingly similar. At this stage, the animals were offered a third button which gave them a small reward regardless of their response. When the two sounds become difficult to discriminate, animals pressed the third button more frequently than the other two. In other words, they demonstrated that they could mobilize their knowledge and pressed the third button only when they had trouble differentiating the frequencies of the sounds. Results like this have since been recorded in monkeys (Morgan *et al.*, 2014), rats (Foote and Crystal, 2007) and even bees (Perry and Barron, 2013). The system could be made more complicated, for example by asking animals to press buttons some time after they had made correct choices in previous tests. So Hampton (2019) found that macaques avoided answering tests of memory when they had forgotten the 'right' answers. Then, they looked for the extra information that they were missing and made consistent 'bets' about the right answer after making judgements. This was clear evidence that they had thought about their thinking and showed that they were capable of a metacognition.

A special aspect of animal consciousness has been studied further when examining relationships between individuals. Some animals must constantly adjust what they do according to what their partners are doing. It is now possible to test hypotheses related to the personal thinking and accumulation of knowledge by the individual, but also its responses to the knowledge of related members of its group. This theory of mind is now well documented in crows (Bugnyar and Heinrich, 2005) and western scrub jays (Dally *et al.*, 2005) which change their strategies for hiding food depending on who is watching them. Call and Tomasello (2008) showed experimentally that chimpanzees understand both the goals and the intentions of other individuals of their own species and humans as well. Other texts report the ability of certain animals, including birds, monkeys and rodents to 'deceive'. An example of this comes from poultry where Gyger *et al.* (1986) demonstrated an 'audience' effect (Fig. 4.1). A rooster in one half of a double cage was shown the silhouette of a raptor. The rooster only sounded an alarm if there were chickens in the other half of the cage. He remained silent if it were empty or contained another species, a quail. Marler *et al.* (1986) also reported the response of roosters to the image of a raptor when placed in a small pen that was surrounded by either chickens or roosters. When there were chickens, the rooster sounded the alarm and sought safety. When in the presence of other roosters, this same rooster sought safety without making a cry. These authors call this behaviour 'deception'.

In some social species there is individual recognition, for example through sound (Lengagne *et al.*, 1999) and sight (Dasser, 1987; Tibbets, 2002; Coulon *et al.*, 2009; Autier-Dérian *et al.*, 2013). In pairs of animals, there is a dominance and subordination relationship. One of the two has priority access to resources like food, environment or sex. Dominance and subordination often lead to a quasi-transitivity in the relationship. If A dominates B and B dominates C, and if A also dominates C, it is called transitivity. But, transitivity like this is not always seen. For example, in a group of Japanese macaques (*Macaca fuscata*) that was organized into maternal lines, researchers showed the importance of alliances between individuals in that context of dominance and subordination (Chapais, 1988). Two individuals of the most subordinate line could dominate a single individual of a more dominant line. Also, in Capuchin monkeys (Perry *et al.*, 2004)

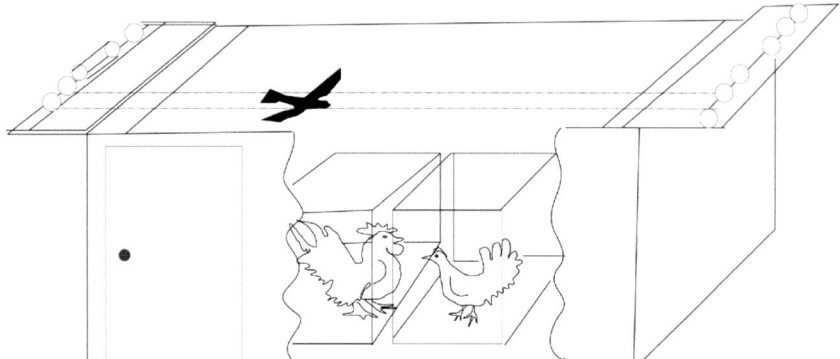

Fig. 4.1. The 'audience effect'. This is demonstrated using the experimental set-up shown: the cage where the rooster is maintained is placed within a room; a device enables a model of a bird of prey to 'fly' over the cage; another cage is placed in front of the rooster's cage. In this second cage, depending on the test situation, either a hen or a quail is placed. In the control situation, this cage is empty. Using this set-up it was found that the rooster only sounded an alarm if there were chickens in the second cage. He was silent if it was empty or contained a quail. In other experimental tests several hens or roosters were placed in front of the rooster. Here the experimental rooster was silent when with other roosters but not with hens. Adapted from Gyger *et al.*, 1986 and from Evans and Marler, 1991.

and hyenas (Engh *et al.*, 2005) two subordinates can come together to win resources against an individual that is dominant to each of them. In other words, subordinates can become allies not only by being aware of each other but by knowing their combined status in a dominance hierarchy.

There are many other examples that show animal species having the capacity to be conscious. Tools have been given as examples of consciousness in non-human primates (Boesch and Boesch, 1983; Whiten *et al.*, 1999; Fragaszy *et al.*, 2004), and birds such as the New Caledonian crow (Von Bayern *et al.*, 2018) are all able to use tools to access food.

4.2.3   Can animals show intention, or ability to plan?

In films and popular literature, we often hear that animals breed because they intend to have offspring or behave in a certain way because they have long-term aspirations. These comments need to be taken cautiously because they suggest that animals are capable of consciously planning to have offspring or other future activities. Tattersall (2002, p. 149) cautioned that, when discussing anatomy, 'New structures do not appear for some reason.' The same applies to behaviour. The proposition of intention confuses two different steps: (i) proximal causes; and (ii) ultimate causes (Gérard *et al.*, 1997). Proximal causes are those that lead animals to perform an act. Ultimate causes are those that infer that performing an act will allow the species to continue to survive and reproduce. But to assume that animals consciously achieve their ultimate expectations is very surprising because, even in humans, this assumption is seldom valid or is at least very weak. To infer that humans have sexual intercourse to ensure the birth of their descendants ignores that carnal desire is a powerful driver of copulation. Certainly, we can infer that humans understand the consequences of mating for the continuation of the species, without necessarily being conscious of this at the time. Such complex partial dissociation between copulation and procreation may be part of reasoning in humans, but it is highly unlikely in animals. Animals try to copulate, the proximal cause, and this allows their species to perpetuate itself, that is the ultimate cause. In fact, new reproductive techniques such as embryo transfer and artificial insemination separate these two causes even further from each other. These techniques are now widely used for domestic species, such as chickens, turkeys, ducks, geese, cattle, sheep, horses and pigs, and interfere with the normal or natural interplay between males and females.

Foreplay or courting behaviour, in both males and females, is complex in all domestic animal species and requires cooperation and acceptance of both sexes. The only possible exception that we can cite are a small percentage (3%) of birds which have a penis, including ducks, geese, swans and large flightless birds such as ostriches and emus (Young, 2013). In other birds and domestic animals, cooperation between males and females is essential to ensure copulation. Such behaviour has been studied in sheep demonstrating the importance of the environment at mating (Inkster, 1957; Lindsay and Robinson, 1961). On this subject, an anecdote was reported to us by a breeder in New Zealand in 1985. One feminist had argued that rams were rapists. That a facetious breeder tied a ram to a stake in a meadow where non-pregnant ewes were roaming freely. At the end of the breeding season all the ewes were pregnant! Anecdotes like this do not make a law, but they demonstrate how anthropocentric thinking can blur our rationality.

The pigeon is the only domestic species in which males and females exhibit analogous behaviour and this, in both sexes, is highly complex. For example, a pigeon breeder who places six males and six females in an aviary may find the next day only three pairs and six unmatched birds. Some singles deliberate for a long time before they finally find partners with whom they will mate. The feeding habits of pigeons also reflect this parallel behaviour which is different to that of all other birds and continues long after mating. The young are fed by both parents, who produce in their crops a specific product called 'crop milk'.

We can also hypothesize intent when studying the behaviour of many species that care for their young. Females, and sometimes males, are said to express behaviour with the intent of protecting their young. Even that assumption can be questioned. Lynch *et al.* (1980) suggested that this may be difficult to infer from the behaviour that he observed in Australian merino lambs. Many of these are born at the end of winter when low temperatures and wind can chill the lambs, sometimes resulting in mortality. To combat the unfavourable climatic conditions, Lynch offered shelters to the ewes, but the ewes only used the shelters if they, themselves, benefited. Lynch *et al.* observed that when half of the ewes were shorn and the other half left in wool, the shorn ewes were, of course, more sensitive to cold than those in wool. This induced them to seek shelter more frequently and as a result more of their lambs survived. So the ewes did show behaviour that favoured the survival of their lambs, which was the ultimate cause, but they were primarily sheltering themselves, which was the proximal cause.

Another example of a possible misinterpretation of intention can be found in the Pyrenean chamois (Gérard *et al.*, 1997) which faces many predators such as wolves, bears and lynx. The young gather in compact groups and the females remain around these 'crèches'. This behaviour was thought to be an anti-predatory strategy, which implied that mothers surrounded their young when they were resting with the intention of protecting them. However, there is an alternative hypothesis that is less complicated and does not involve the intention of the mothers. That behaviour is justified by three simple tenets: (i) young want to be close to other young; (ii) mothers want to be close to their own young; and (iii) mothers are not attracted to other mothers. These tenets explain the behavioural structure in the chamois without invoking an 'anti-predatory strategy'. Thus, Gérard *et al.* proposed that grouping is a proximal cause, even if the ultimate cause is in fact an anti-predatory strategy.

Nevertheless, the term, 'intention' was used by early ethologists like Tinbergen to explain locomotory behaviour that was separate from its primary function of just moving about. He incorporated it into complex sequences of behaviour that were used for communication. Communication signals of this sort have probably become, in the course of evolution, stereotyped and, like 'intention' movements, they are incomplete. Tinbergen (1952) called this process, ritualization. The term 'intention' when used in this sense therefore has no finalist connotation, but only refers to the incomplete nature of the movement. This is clearly illustrated if we consider the aggressive behaviour of mammals and birds. Aggressive behaviour initially involves visually confronting the protagonist. Then, in a controlled progression, subsequent behaviour of the aggressor involves moving forward and displaying strong muscular tension. This locomotory activity towards the protagonist is clearly an intention to attack which is unfinished (Darwin, 1872; Deputte, 2007).

Apart from stereotyped responses to stimuli, animals show that they can also memorize events, which they may or may not evoke when expressing subsequent behaviour. This can result in complex cognitive action to produce responses tailored to the situations they face. In other words, they deliberate about their own actions, which we defined earlier as consciousness.

Emotions are essential elements in the consciousness of animals. They indicate that the animal can appraise its own feelings and the intricacy of the situation in which it finds itself. Emotions characterize the meaning that animals give to the events they have to experience, and so are much more than simple mechanical responses to stimuli. Therefore, we conclude that proximal causes should not be confused with ultimate causes.

References

Autier-Dérian, D., Deputte, B.L., Chalvet-Monfray, K., Coulon, M. and Mounier, L. (2013) Visual discrimination of species in dogs (*Canis familiaris*). *Animal Cognition* 16(4), 637–651. DOI: 10.1007/s10071-013-0600-8.

Boesch, C. and Boesch, H. (1983) Optimisation of nut-cracking with natural hammers by wild chimpanzees. *Behaviour* 83(3–4), 265–286. DOI: 10.1163/156853983X00192.

Boissy, A., Manteuffel, G., Jensen, M.B., Moe, R.O., Spruijt, B. *et al.* (2007) Assessment of positive emotions in animals to improve their welfare. *Physiology & Behavior* 92(3), 375–397. DOI: 10.1016/j.physbeh.2007.02.003.

Bugnyar, T. and Heinrich, B. (2005) Ravens, *Corvus corax*, differentiate between knowledgeable and ignorant competitors. *Proceedings of the Royal Society B: Biological Sciences* 272(1573), 1641–1646. DOI: 10.1098/rspb.2005.3144.

Byrne, R.W. and Whiten, A. (1988) *Machiavellian Intelligence: Social Expertise and the Evolution of Intellect in Monkeys, Apes and Humans.* Clarendon Press, Oxford.

Call, J. and Tomasello, M. (2008) Does the chimpanzee have a theory of mind? 30 years later. *Trends in Cognitive Sciences* 12(5), 187–192. DOI: 10.1016/j.tics.2008.02.010.

Chapais, B. (1988) Experimental matrilineal inheritance of rank in female Japanese macaques. *Animal Behaviour* 36(4), 1025–1037. DOI: 10.1016/S0003-3472(88)80062-9.

Clayton, N.S. and Dickinson, A. (1998) Episodic-like memory during cache recovery by scrub jays. *Nature* 395, 272–274. DOI: 10.1038/26216.

Coulon, M., Deputte, B.L., Heyman, Y. and Baudoin, C. (2009) Individual recognition in domestic cattle (*Bos taurus*): evidence from 2D-images of heads from different breeds. *PloS ONE* 4(2), e4441. DOI: 10.1371/journal.pone.0004441.

Crystal, J.D. (2010) Episodic-like memory in animals. *Behavioural Brain Research* 215(2), 235–243. DOI: 10.1016/j.bbr.2010.03.005.

Dally, J.M., Emery, N.J. and Clayton, N.S. (2005) Cache protection strategies by western scrub-jays, *Aphelocoma californica*: implications for social cognition. *Animal Behaviour* 70(6), 1251–1263. DOI: 10.1016/j.anbehav.2005.02.009.

Damasio, A. (1999) *The Feeling of What Happens: Body and Emotion in the Making of Consciousness.* Harcourt Brace & Co, San Diego, California, 424 pp.

Dantzer, R. and Mormède, P. (1979) *Le Stress en Élevage Intensif.* Actualités Scientifiques et Agronomiques de l'INRA, Paris, 117 pp.

Darwin, C. (1872) *The Expression of the Emotions in Man and Animals.* John Murray, London. DOI: 10.1037/10001-000.

Dasser, V. (1987) Slides of group members as representations of the real animals (*Macaca fascicularis*). *Ethology: Formerly Zeitschrift Fur Tierpsychologie* 76(1), 65–73. DOI: 10.1111/j.1439-0310.1987.tb00672.x.

Deputte, B.L. (2007) Comportements d'agression chez les vertébrés supérieurs, notamment chez le chien domestique (*Canis familiaris*). *Bulletin de l'Académie Vétérinaire de France* 60, 349–358.

Destrez, A., Deiss, V., Lévy, F., Calandreau, L., Lee, C. *et al.* (2013) Chronic stress induces pessimistic-like judgment and learning deficits in sheep. *Applied Animal Behaviour Science* 148(1–2), 28–36. DOI: 10.1016/j.applanim.2013.07.016.

de Waal, F. (1984) *Chimpanzee Politics: Power and Sex among Apes.* HarperCollins, New York.

de Waal, F. (2016) *Are We Smart Enough to Know How Smart Animals Are?* Granta, London, 340 pp.

Descartes, R. (1637) *Discours de la Méthode.* Librairie philosophique, J. Vrin, Paris.

Destrez, A., Deiss, V., Leterrier, C., Calandreau, L. and Boissy, A. (2014) Repeated exposure to positive events induces optimistic-like judgment and enhances fearfulness in chronically stressed sheep. *Applied Animal Behaviour Science* 154, 30–38. DOI: 10.1016/j.applanim.2014.01.005.

Doré, F.Y. and Mercier, P. (1999) *Les Fondements de l'Apprentissage et de la Cognition.* Gaétan Morin, Quebec, Canada, 514 pp.

Douglas, C., Bateson, M., Walsh, C., Bédué, A. and Edwards, S.A. (2012) Environmental enrichment induces optimistic cognitive biases in pigs. *Applied Animal Behaviour Science* 139(1–2), 65–73. DOI: 10.1016/j.applanim.2012.02.018.

Eacott, M.J. and Easton, A. (2012) Remembering the past and thinking about the future: is it really about time? *Learning and Motivation* 43(4), 200–208. DOI: 10.1016/j.lmot.2012.05.012.

Eacott, M.J., Easton, A. and Zinkivskay, A. (2005) Recollection in an episodic-like memory task in the rat. *Learning and Memory* 12(3), 221–223. DOI: 10.1101/lm.92505.

Engh, A.L., Siebert, E.R., Greenberg, D.A. and Holekamp, K.E. (2005) Patterns of alliance formation and postconflict aggression indicate spotted hyaenas recognize third-party relationships. *Animal Behaviour* 69(1), 209–217. DOI: 10.1016/j.anbehav.2004.04.013.

Evans, C.S. and Marler, P. (1991) On the use of video images as social stimuli in birds: audience effects on alarm calling. *Animal Behaviour* 41(1), 17–26. DOI: 10.1016/S0003-3472(05)80499-3.

Fabre-Nys, C. and Venier, G. (1987) Development and use of a method for quantifying female sexual behaviour in ewes. *Applied Animal Behaviour Science* 17(3–4), 289–304. DOI: 10.1016/0168-1591(87)90153-5.

Ferkin, M.H., Combs, A., del Barco-Trillo, J., Pierce, A.A. and Franklin, S. (2008) Meadow voles, *Microtus pennsylvanicus*, have the capacity to recall the 'what', 'where', and 'when' of a single past event. *Animal Cognition* 11(1), 147–159.

Foote, A.L. and Crystal, J.D. (2007) Metacognition in the rat. *Current Biology* 17, 551–555. DOI: 10.1016/j.cub.2007.01.061.

Fragaszy, D., Izar, P., Visalberghi, E., Ottoni, E.B. and de Oliveira, M.G. (2004) Wild capuchin monkeys (*Cebus libidinosus*) use anvils and stone pounding tools. *American Journal of Primatology* 64(4), 359–366. DOI: 10.1002/ajp.20085.

Gérard, J.-F., Gonzalez, G., Guilhem, C., LePendu, Y., Quenette, P.Y. *et al.* (1997) Emergences collectives chez les ongulés sauvages. In: Theraulaz, G. and Spitz, F. (eds) *Auto-Organisation et Comportement.* Hermès, Paris, pp. 171–186.

Gyger, M., Karakashian, S.J. and Marler, P. (1986) Avian alarm calling: is there an audience effect? *Animal Behaviour* 34(5), 1570–1572. DOI: 10.1016/S0003-3472(86)80229-9.

Hampton, R.R. (2019) Monkey metacognition could generate more insight. *Animal Behavior and Cognition* 6(4), 230–235. DOI: 10.26451/abc.06.04.02.2019.

Inkster, I.J. (1957) The mating behaviour of sheep. In: *Sheep Farming Annual.* Massey Agricultural College, Auckland, New Zealand, pp. 163–169.

Jelbert, S.A., Hurly, T.A., Marshall, R.E.S. and Healy, S.D. (2014) Wild, free-living hummingbirds can learn what happened, where and in which context. *Animal Behaviour* 89, 185–189. DOI: 10.1016/j.anbehav.2013.12.028.

Keverne, E.B., Levy, F., Poindron, P. and Lindsay, D.R. (1983) Vaginal stimulation: an important determinant of maternal bonding in sheep. Science 219(4580), 81–83. DOI: 10.1126/science.6849123.

Kouwenberg, A.L., Walsh, C.J., Morgan, B.E. and Martin, G.M. (2009) Episodic-like memory in crossbred Yucatan minipigs (*Sus scrofa*). *Applied Animal Behaviour Science* 117(3–4), 165–172. DOI: 10.1016/j.applanim.2009.01.005.

Laureys, S., Owen, A.M. and Schiff, N.D. (2004) Brain function in coma, vegetative state, and related disorders. *The Lancet. Neurology* 3(9), 537–546. DOI: 10.1016/S1474-4422(04)00852-X.

Le Neindre, P., Bernard, E., Boissy, A., Boivin, X., Calandreau, L. *et al.* (2017) *Animal Consciousness.* EFSA (European Food Safety Authority) supporting publication, Parma, Italy, 165 pp.

Lengagne, T., Jouventin, P. and Aubin, T. (1999) Finding one's mate in a king penguin colony: efficiency of acoustic communication. *Behaviour* 136(7), 833–846. DOI: 10.1163/156853999501595.

Le Ny, J.F. (2002) Cognition. In: Tiberghien, G.S. (ed.) *Dictionnaire des Sciences Cognitives.* Armand Colin, Paris, pp. 70–71.

Levy, F., Poindron, P. and Le Neindre, P. (1983) Attraction and repulsion by amniotic fluids and their olfactory control in the ewe around parturition. *Physiology & Behavior* 31(5), 687–692. DOI: 10.1016/S0031-9384(83)80004-3.

Lindsay, D.R. and Robinson, T.J. (1961) Studies on the efficiency of mating in the sheep II. The effect of freedom of rams, paddock size, and age of ewes. *The Journal of Agricultural Science* 57(1), 141–145. DOI: 10.1017/S0021859600050127.

Low, P., Panksepp, J., Reiss, D., Edelman, D., Van Swinderen, B. *et al.* (2012) The Cambridge declaration on consciousness. In: *Proceedings of the Francis Crick Memorial Conference on Consciousness in Human and Non-human Animals,* University of Cambridge, pp. 1–2.

Lynch, J.J., Mottershead, B.E. and Alexander, G. (1980) Sheltering behaviour and lamb mortality amongst shorn Merino ewes lambing in paddocks with a restricted area of shelter or no shelter. *Applied Animal Ethology* 6(2), 163–174. DOI: 10.1016/0304-3762(80)90067-X.

Marler, P., Dufty, A. and Pickert, R.L. (1986) Vocal communication in the domestic chicken: II. Is a sender sensitive to the presence and nature of a receiver? *Animal Behaviour* 34, 194–198. DOI: 10.1016/0003-3472(86)90023-0.

Menault, E. (1868) *L'intelligence des Animaux.* Librairie de L'Hachette, Paris, 343 pp.

Miklosi, A. (2015) *Dog Behaviour, Evolution and Cognition.* Oxford University Press, Oxford, 274 pp.

Morgan, G., Kornell, N., Kornblum, T. and Terrace, H.S. (2014) Retrospective and prospective metacognitive judgments in rhesus macaques (*Macaca mulatta*). *Animal Cognition* 17(2), 249–257. DOI: 10.1007/s10071-013-0657-4.

Nicolas, A. (2020) *L'imposture Antispéciste*. Desclée de Brouwer, Bilbao, Spain, 263 pp.

Pena-Guzman, D. (2022) *When Animals Dream: The Hidden World of Animal Consciousness*. Princeton University Press, Princeton, New Jersey, 272 pp.

Perry, S., Barrett, H.C. and Manson, J.H. (2004) White-faced capuchin monkeys show triadic awareness in their choice of allies. *Animal Behaviour* 67(1), 165–170. DOI: 10.1016/j.anbehav.2003.04.005.

Perry, C. and Barron, A. (2013) Honeybees selectively avoid difficult choices. *Proceedings of the National Academy of Sciences of the United States of America* 110(47), 19155–19159.

Poindron, P. and Le Neindre, P. (1980) Endocrine and sensory regulation of maternal behaviour in the ewe. *Advances in the Study of Behavior* 11, 75–119.

Poindron, P., Levy, F., Le Neindre, P. and Keverne, E.B. (1986) The roles of genital stimulation, oestrogens and olfaction in the maternal bonding of sheep and other mammals. *Hormones and Behavior* 20, 538–547.

Poindron, P., Le Neindre, P., Lèvy, F. and Keverne, E.B. (1984) Les mécanismes physiologiques de l'acceptation du nouveau-né chez la brebis. *Biologie du Comportement* 9, 65–88.

Premack, D. and Woodruff, G. (1978) Does the chimpanzee have a theory of mind? *Behavioral and Brain Sciences* 1(4), 515–526. DOI: 10.1017/S0140525X00076512.

Prunier, A. (1989) Influence de la présentation au verrat sur l'âge à la puberté des truies. *INRAE Productions Animales* 2(1), 65–72. DOI: 10.20870/productions-animales.1989.2.1.4401.

Raby, C.R. and Clayton, N.S. (2009) Prospective cognition in animals. *Behavioural Processes* 80(3), 314–324. DOI: 10.1016/j.beproc.2008.12.005.

Raby, C.R., Alexis, D.M., Dickinson, A. and Clayton, N.S. (2007) Planning for the future by western scrub-jays. *Nature* 445(7130), 919–921. DOI: 10.1038/nature05575.

Roberts, W.A. (2002) Are animals stuck in time? *Psychological Bulletin* 128(3), 473–489. DOI: 10.1037/0033-2909.128.3.473.

Roberts, W.A. (2007) Mental time travel: animals anticipate the future. *Current Biology* 17(11), R418–R420. DOI: 10.1016/j.cub.2007.04.010.

Rosenblatt, J., Siegel, H. and Mayer, A. (1979) Progress in the study of maternal behaviour in the rat: hormonal, non-hormonal, sensory, and developmental aspects. *Advances in the Study of Behavior* 10, 225–311.

Signoret, J.P. (1970a) Reproductive behaviour of pigs. *Journal of Reproduction and Fertility* 11, 105–117.

Signoret, J.P. (1970b) Sexual behaviour patterns in female domestic pigs (*Sus scrofa* L.) reared in isolation from males. *Animal Behaviour* 18, 165–168. DOI: 10.1016/0003-3472(70)90086-2.

Signoret, J.P. (1975) The behaviour of swine. In: Hafez, E.S.E. (ed.) *Behaviour of Domestic Animals*. Baillière Tindall, London, pp. 295–329.

Smith, J.D., Schull, J., Strote, J., McGee, K., Egnor, R. *et al.* (1995) The uncertain response in the bottlenosed dolphin (*Tursiops truncatus*). *Journal of Experimental Psychology. General* 124(4), 391–408. DOI: 10.1037//0096-3445.124.4.391.

Soley, F.G. and Alvarado-Díaz, I. (2011) Prospective thinking in a mustelid? *Eira barbara* (Carnivora) cache unripe fruits to consume them once ripened. *Naturwissenschaften* 98(8), 693–698. DOI: 10.1007/s00114-011-0821-0.

Streri, A., Lhote, M. and Dutilleul, S. (2000) Haptic perception in newborns. *Developmental Science* 3(3), 319–327. DOI: 10.1111/1467-7687.00126.

Süddendorf, T. and Busby, J. (2003) Like it or not? the mental time travel debate: reply to clayton et al. *Trends in Cognitive Sciences* 7, 437–438.

Tattersall, I. (2002) *The Monkey in the Mirror*. Oxford University Press, Oxford, UK, 220 pp.

Thimonier, J., Cognié, Y., Lassoued, N. and Khaldi, G. (2000) L'effet mâle chez les ovins: une technique actuelle de maîtrise de la reproduction. *INRA Productions Animales* 13, 223–231.

Tibbets, E.A. (2002) Visual signals of individual identity in the wasp *Polistes fuscatus*. *Proceedings of the Royal Society B: Biological Sciences* 269, 1423–1428.

Tinbergen, N. (1952) Derived activities: their causation, biological significance, origin. *The Quarterly Review of Biology* 27, 1–32.

Tinbergen, N. (1963) On aims and methods of ethology. *Zeitschrift Für Tierpsychologie* 20(4), 410–433. DOI: 10.1111/j.1439-0310.1963.tb01161.x.

Vauclair, J. (1992) *L'intelligence de l'Animal*. Éditions du Seuil, Paris, 238 pp.

Von Bayern, A., Danel, S., Auersperg, A.M.I., Mioduszewska, B. and Kacelnik, A. (2018) Compound tool construction by New Caledonian crows. *Scientific Reports* 8, 15676.

Whiten, A., Goodall, J., McGrew, W.C., Nishida, T., Reynolds, V. *et al.* (1999) Cultures in chimpanzees. *Nature* 399(6737), 682–685. DOI: 10.1038/21415.

Young, E. (2013) How chickens lost their penises (and ducks kept theirs). *National Geographic*, 6 June. Available at: https://www.nationalgeographic.com/scienc e/article/how-chickens-lost-their-penises-ducks-kept-theirs (accessed 14 June 2022).

Zinkivskay, A., Nazir, F. and Smulders, T.V. (2008) What-where-when memory in magpies (*Pica pica*). *Animal Cognition* 12(1), 119–125. DOI: 10.1007/ s10071-008-0176-x.

The Human Factor: How Do Animals Adapt To Humans?

5

5.1 'Natural' Living Conditions

In broad terms, people who legislate want domestic animals and pets to be raised under 'natural conditions' and this puzzles some biologists. For example, the British Farm Animal Welfare Committee (FAWC; previously the Farm Animal Welfare Council) proposes a list of freedoms for animal welfare among which is the freedom to express 'normal' behaviour (FAWC, 1979). These legislative requirements assume that certain conditions can be defined as 'natural' which infers those other conditions are 'not natural' or are 'artificial'. For us, this is far too simplistic and superficial given the complexity and variety of natural conditions and the adaptation of animals to the environment they share with humans. In this chapter, we will look at this concept.

First, let us be clear that a population of animals living in an environment does not mean that these animals are more adapted than other animals to that environment – or that they might not be better adapted if they were in another environment. If we define adaptation as the ability to survive and have offspring, then the reasoning 'they are adapted because they are there' is ultimately circular. Nevertheless, this reasoning is often invoked when discussing both wild animals and domestic animals. One could turn the argument around by arguing that they are there because they are adapted.

Every species of animal adapts so that it can live and survive in its environment. This is so for the clam on the Brittany beaches where it flourishes in alternative phases of immersion at high and low tides (Le Gros, 2017). It is equally so for the wingless flies of the Austral Islands near Antarctica. These flies store fat in the space where other flies have wings so that they are not carried away by the constant Antarctic winds, but resist fasting in periods of extreme cold because of its energy reserves of fat (Vernon, 1986; Vernon and Vannier, 1996). Adaptation mechanisms are multiple and sometimes very surprising.

© CAB International 2022. *Interacting with Animals: Understanding their Behaviour and Welfare* (Pierre Le Neindre and Bertrand L. Deputte) DOI: 10.1079/9781800622418.0005

Almost all non-human animal species, individually or in groups, adapt to the environment rather than adapting the environment to fit them, and seldom destroy or overexploit it. However, there are some examples of destruction not triggered by humans. For example, the disappearance in the Pleistocene of the megafauna of South America was explained by the junction of the two Americas. The fauna from North America invaded South America and wiped out the fauna in the south (Sánchez Chillon *et al.*, 2003). But there are examples of populations finding ingenious strategies for coping. One striking example is the Australian black kite. These birds can carry flaming fire sticks and drop them to create bushfires which disturb insects that they can then harvest (Bonta *et al.*, 2017).

Calling animals 'domestic' implies that they live in the same environment as humans. Thus, mice and sparrows are technically domestic. Understanding this domestication helps us to analyse the animals with which we are in close contact and to understand how they may continue to adapt. We will ignore mice and sparrows here, and look at domestication in the more limited sense, where domestic animals are those chosen by humans to live with them in a more or less common environment. Studying such animals has the advantage that we understand their man-made environment better than we understand most natural environments. Moreover, the animals have developed their response to it over a relatively short time and these responses can be reproduced, making them easier to study scientifically. For example, Price (1984) found profound changes in the reproductive behaviour of gerbils that had been taken from the wild and then kept for several generations in a rearing environment.

Some of the well-known originators of the science of ethology such as Lorenz (1957) described domestication of animals as a form of degeneration, as he claimed that it often resulted in abnormal physical and behavioural changes. It is probably more accurate to describe this evolution as an adaptation to constraints imposed by humans. In fact, in most animals it has resulted in a significant increase in reproductive success in farm environments, so that poultry, sheep, goats, pigs and cattle, among others, have proliferated significantly better compared with the populations of origin. And, even without the influence of humans, in so-called wild or natural environments, animals are constantly adapting to the many biotic or abiotic constraints of these environments. However, one difference is the time scale. Adaptation is extremely short in the case of artificial selection and quite long in the wild where natural selection takes place. Another difference is that humans have often set goals for change and have used new reproductive techniques including modifying DNA, cloning and *in vitro* fertilization (IVF). Adaptation may be physiological, such as being able to change a diet or to successfully metabolize products that are toxic to other animals, or it may be behavioural. The distinctive role of domestication is that it identifies humans as the driving force behind adaptation, through their use of artificial selection.

Over time, the differentiation of wild species came about by natural selection. This was not only by chance or by geographical isolation but also by their adapting to constraints that were particular to the environments in which they lived. By contrast, domestication has, so far, not led to differentiation as extreme as creating new species. Even though some animals have been domesticated for a long time on an historical scale, all domesticated species are still able to reproduce with individuals from non-domesticated populations of their same species.

None the less, within domesticated species, we recognize different populations, or breeds. These breeds are distinguished not only by their physical appearance, weight or productivity, but also by their behaviour. Each breed is the result of conscious, unconscious or fortuitous selection by breeders to obtain animals corresponding to what they want. For example, in cattle there are two major species, *Bos taurus* and *Bos indicus*. *B. taurus* inhabit temperate and cold zones, *B. indicus* inhabit warm zones.

Within temperate zones, we can distinguish three types of domestic cattle, often associated with their breeding environments:

1. Suckler cows of 'meat' breeds that feed their calves for several months and are not milked by humans. In France, the most common breeds in this category are the Charolais and the Limousin.
2. Cows of dairy breeds, such as Holstein, that are separated from their calves soon after parturition and are milked by humans throughout their lactation.
3. Dual-purpose breeds that are usually raised in harsh environments while producing milk for humans and calves, often in lower quantities than specialist breeds. They are, or were, milked while continuing to suckle their calves.

Breeders selected these breeds which, as a result, differ not only in their physical characteristics but also in their behaviour. The great difference in behaviour is illustrated in a study that compared a dairy breed, the Friesian, or Holstein, that produces copious amounts of milk, and a dual-purpose, hardy breed, the Salers that produces less milk but often does so in harsh environments (Le Neindre, 1989a, b). On dairy farms, farmers seeking as much milk as possible wean calves from their mothers as early as possible and limit contact between calves and their mothers by feeding milk to the calf from a bucket. To consume this milk from the bucket, calves must learn to drink the milk and no longer suck as they do from the udder or a bottle. Calves of both breeds suck from their mothers without problems. But Salers calves find it much harder than Friesians to learn to drink from a bucket. This seemingly simple act of drinking a liquid is therefore not spontaneous in young calves but a learned trait for which breeders can select.

In females of non-domesticated breeds or species, milk ejection, or 'let-down', is initiated when the young attempts to suck (Labussière and

Fig. 5.1. Egyptian painting of parturition of a cow and the first interaction between the cow and its calf. During these two events, a human is present either to help the parturition or milk the cow in presence of the new-born calf. Egypt – Metchetchi tomb – 2350 BC; Louvre Museum, Paris. Photo courtesy of B.L. Deputte.

Richard, 1966). For centuries, farmers 'stole' milk from cows, by letting the calf suck for a short time to trigger 'let-down' before milking them. The calves were then allowed to suck again, which sustained them with the extra milk and also prolonged lactation. Prehistoric and pharaonic rock carvings and paintings illustrate this procedure (Fig. 5.1). This technique is still used in many African and Asian countries with lactating females of cattle and other species, especially mares and yaks.

The milk that cows produce is generally intended for their calves. In fact, calves in a herd of suckler cows can only suck from their own mothers because cows do not allow any other calf but their own to suckle. If a Salers cow dies, its calf will probably starve to death because other cows in the herd deny it the milk it needs for survival unless the farmer intervenes. Friesian cows on the other hand, are much less selective than Salers and allow foreign calves to suck from them relatively easily. We postulate that this difference is the result of thousands of years of selection.

However, milk production is only one of many traits for which breeders deliberately select. They have also found it beneficial to select behavioural characteristics, and in particular social behaviour. Dairy calves are usually raised in groups, often in confinement and the composition of these groups changes frequently. In contrast beef calves live continually with their mothers outdoors, at least during the grazing season. The difference in rearing environment has been shown by Le Neindre (1989a) to have a pronounced effect on their subsequent social behaviour. He observed Salers and Friesian calves in two early rearing systems: (i) suckling their mothers (Salers); and (ii) artificial, or bucket, rearing (Friesians). After weaning, they continued to be reared together in groups. In these post-weaning groups, most of the Salers dominated Friesians. The only exceptions were those Salers that had been raised without their mothers. By contrast the Friesians' mode of rearing at a young age did not affect their subsequent social behaviour. Indeed, they could be considered to be asocial.

Lorenz (1957) hypothesized that asocial animals are better suited to intensive farming than social animals. He wrote that 'the decrease in social instincts and inhibitions is extremely useful in the battle of modern competition, and this is how unsocial or even asocial beings are far more successful than the super-active animals' (Lorenz, 1957, p. 462). Mills *et al.* (1995) working on quail, selected birds with very high, or very low levels of sociality and came to the same conclusion.

This raises the concept of hardiness which we should consider more closely. Unfortunately, the term 'hardy' has many interpretations. It is generally used by breeders to highlight the good features of animals. For example, breeders of Salers and Holsteins both claim that their animals are 'hardy'. For Salers, this means that they live and continue to reproduce when challenged with high or low temperatures, poor quality feed and rugged terrain. The living conditions of Salers have always been difficult, not only because they graze in mountainous regions but also because they endure challenging winters. For Holsteins, hardiness is the ability to adapt to intensive livestock environments and to live in unstable breeding groups. So, logically, they conclude that Holsteins adapt to all the constraints imposed on them in dairies and, therefore, they can be qualified as hardy. So they are well adapted to the environment imposed on them. Therefore 'hardiness' is a very poor concept unless it is defined more clearly.

In other species, other traits have been modified by selection both by the environment and by breeders. This has resulted in females of domestic strains now having different reproductive traits from those of their ancestors. For example, silver foxes bred in captivity ovulate many times per year, which makes breeding them more profitable while foxes in the wild have only one ovulation per year (Belyaev *et al.*, 1985). Similarly, laying hens have been intensively selected to produce many eggs. This high production is accomplished in part by limiting brooding behaviour. Hens of older or unselected strains display brooding behaviour soon after laying their eggs. They need to make nests, incubate the eggs and care for the young. Modern strains show little or none of these behaviours to the extent that reproduction is now only possible thanks to artificial brooding (Sauveur, 1988).

In reality, deliberately modified behavioural characteristics are found in virtually all species of domestic animals. However, very few of these modifications could be classed as all-or-nothing. Typically, a modification may result in a character being attenuated rather than eliminated or appearing only in a particular context. Moreover, adaptations to meet the limitations of domestic environments often have their counterparts in populations of wild animals that have adapted to new limitations in the wild. This is especially the case in environments where the species have been introduced relatively recently to a new area. A full-scale case in point is in Australia. In this island continent, many species of animals that were imported had already been domesticated. Occasionally, their descendants escaped the domestic environment and became feral. So cats, for example, are now widespread. They have developed a phenotype that is larger in size than domestic cats and characterized by their being very difficult to tame. They thrive by destroying native wildlife, in some cases to the point of extinction for many species of marsupials and birds (Woinarski *et al.*, 2019). Similarly, brumbies, or feral horses, thrive in the bush but they are difficult to tame and therefore considered unsuitable for re-establishing as domestic animals. Many populations have also been introduced to smaller isolated islands. A population originating from five cattle imported in the 19th century to the island of Amsterdam in the southern Indian Ocean also flourished before it was recently exterminated because of its deleterious effect on native birds. When observed in 1969 the social organization of these animals was probably close to that of the original wild European population (Lésel, 1969).

To adapt to new environments animals have developed specific features. For example, North Ronaldsay sheep on the Shetland Islands have adopted a unique diet. They consume seaweed which is the only foodstuff available. As a consequence, they can no longer safely eat pasture grass in the meadows. They can tolerate a high amount of salt in their diet but are sensitive to copper levels that do not trouble other sheep (Hansen *et al.*, 2003).

Surprisingly, only a few domesticated species seem not to have the capacity to return to the wild. For example, we have seen that modern breeds of poultry no longer display brooding behaviour which would be essential in the wild. Extreme-type animals, whether kept for recreation or for production, would also find it difficult to live without the support of humans. For example, short-headed breeds of dogs suffer from respiratory problems (Denis, 2007). Double-muscled cattle such as the Belgian Blue breed (also known as Belgian White Blue) have difficulties when calving, often requiring intervention by humans to prevent death of the calf, the mother, or both (Vissac *et al.*, 1974).

Selection by breeders can have very important ethical consequences. The human species relies heavily on technology and depends on a type of understanding that is different from non-human species – a consequence of having language with which to communicate – and its caring for the environment seems to have become unimportant. Consequently, overexploitation of both the environment and the animals within it is all too common. One might have thought that the cognitive abilities, including logical reasoning and morality, on which the human species relies would have led it to adapt, like other species that have evolved over geological time, and conserve environments globally for survival of the species. However, humans use their power to eliminate entire populations of their own species, to distance themselves temporarily – on a time scale of a few human generations – from the restrictions they see around them, to interfere with the whole world, and to exhaust the biotic and abiotic resources of the planet for immaterial values such as money. Is this perhaps the true characteristic of humans, the perversion of culture?

So, there are no natural environments per se. There are environments to which the individuals of the animal population concerned have adapted, and in fact are continuing to adapt, that is to live with as few problems as possible, and in which they can reproduce.

5.2   How Do Animals Live With Humans?

This section deals with the consequences for animals that coexist with humans, the conditions under which this coexistence may have existed in the past, and the changes that have resulted.

For thousands of years, animals have lived in association with humans. They share a mutual benefit for their protection, habitats and food, or for cultural or symbolic reasons. We refer to species such as cattle, horses, sheep, goats, reindeer, poultry and cats as domestic. This domestic coexistence was only possible for animals that could withstand the limits that humans imposed on them. Of course, other animals like mice and rats also live in close proximity to humans, but their reproduction does not depend on humans. In this category, too, are some snakes in Asia and genets in Europe. It was once thought that the cat was also in this category.

Humans may not control the entire life cycle of animals, but most depend on humans at some time or another of their lives. For example, working elephants are usually caught from wild populations at a subadult age, and then tamed as work animals. This is also the case for raptors intended for falconry. Every year, hawks and falcons are captured during their migratory passage through Tunisia. At the end of the season, the surviving raptors are released. In Asia, cormorants are used for fishing. Australian Aborigines captured dingoes, which were probably originally Indian wolves, when they were still pups in their burrows, and then tamed them to live with the human tribe and to aid in hunting game animals. Many animals are also kept by humans as mascots, for the pleasure of having them as company. This has been documented especially in Latin America (Digard, 1992).

In the 19th century Isidore Geoffroy Saint-Hilaire founded the *Jardin d'Acclimatation* in Paris to improve the range of animal protein in diets for his fellow citizens. He categorized four stages in the modification of 'useful animals' for the benefit of humans (Geoffroy Saint-Hilaire, 1861, p. 158):

1. *Acclimatization*: imprinting modifications in the animal that make it fit to live and perpetuate its species in its new environment.

2. *Naturalization*: getting the animal to live in a foreign environment like other species that live naturally in that environment.

3. *Taming*: making the animal become familiar with humans and more or less freely accept the constraints imposed by them. Almost all species can be tamed in this way, at least to some extent. The taming process can be complex and is often not irreversible. This is the case of 'haggard' hawks that are captured just as they leave the nest but easily return to their natural environment later. Many 'less intelligent' birds taken from the nest depend on humans for the rest of their lives (Boyer and Planiol, 1948).

4. *Domestication*: inducing animals to become accustomed to living and reproducing in close proximity to humans.

The domestication of the dog is a favourite topic of research. Modern research refutes Darwin's and Lorenz's hypothesis that the dog (*Canis familiaris*) originated from at least two species of canids, the wolf and the coyote or jackal. Domestication of the dog can be looked at in two ways: (i) phylogenetically; and (ii) behaviourally. The phylogenetic part is explained fundamentally by genetics (Wayne and Ostrander, 2007; Frantz *et al.*, 2016); the behavioural part is essentially speculative, in the absence of direct evidence (Clutton-Brock, 1999; Coppinger and Coppinger, 2002; Denis, 2003). Biological reviews on the different species have been published on the subject by Price (1984, 1999), Hemmer (1990), Faure and Le Neindre (2009), while Digard (1988) produced sociological or anthropological studies. All these authors highlight the complexity of analysing the essence of domestication, as we will see below.

Taming and domestication imply a special proximity between animals and humans, at the level of either the individual or the species, particularly when humans interfere with reproduction. Animals and humans are therefore complementary, and taming may precede domestication. Tame tigers, for example, would not be considered domesticated, and taming may be necessary even for animals considered to be domestic.

Animals from domestic populations are maintained and bred without contact with the original population. Some are kept for food or for work. This work can have many forms. It may be physical, by horses and cattle, security by dogs, hunting by dogs, cats and ferrets, and helping people with disabilities by dogs and brown capuchin monkeys (Deputte and Busnel, 1997). And, some animals are kept as mascots, like Australian parakeets that have evolved successfully in captivity in the confined environment of cages.

Commensalism, or where one organism benefits and the other derives neither benefit nor harm, between human and animal can lead to domestication. This two-way association with reciprocal benefit – which resembles symbiosis – is seen regularly between two species that share the same resources without competition. This was described by Isidore

Geoffroy Saint-Hilaire (1861, p. 152): animals that spontaneously settle in our homes or our neighbourhood, that become our guests without our invitation or even against our will and our interests, would, broadly speaking, be domestic animals. This is particularly the case for the two domestic carnivores, dogs and cats, which have become companions to humans in many parts of the world.

Historically, humans chose animals, first empirically and then more systematically, by selecting breeding stock on the basis of their morphology, physiology, physical attributes, colour or even their behaviour. By joining males and females with these attributes, humans hoped that the offspring would have the traits that they were seeking. These offspring were then retained and used to breed the next generation. This type of breeding programme is called artificial selection because it is the result of intentional interference by humans, and not the result of chance and pressures from the natural environment. Artificial selection has been intensified recently by the advent of genetic modification in which direct transformation of genes supposed to be responsible for the selected traits are targeted, either by eliminating them – knock-out individuals – or by ensuring that they are expressed. This artificial selection can make measurable changes over just a few generations, which is a very short time when compared with evolution.

When we contrast domesticated and non-domesticated animals, we find that it is not an all-or-none phenomenon. The relationship between animals and humans varies widely and domestic animals can breed with animals from non-domesticated populations. This happens when breeders reintroduce wild animals into a domestic population when they want to limit the negative effects of inbreeding. For example, the albino ferret (*Mustela putorius furo*) had been bred for two millennia, with the European polecat (*Mustela putorius putorius*) to create commercial crossbred ferrets. Breeders in Brittany let domestic rabbits run free to be mated by wild cottontail rabbits and then raised their offspring. Less anecdotally, reintroduction has been described in Australia where breeders of domestic dogs have mated them with dingoes to obtain animals with characteristics that they considered desirable. In Tibet, herders deliberately mate domestic yak females, which are small, with wild males, which are much larger, to increase the size of their animals (Leslie and Schaller, 2009).

Therefore, artificial selection results in animals with characteristics different from those of the original population but the most consistent characteristic is that the animals live close to humans. In silver foxes Coppinger and Coppinger (2002) proposed that many individuals have become so genetically docile and tolerant that they are even attracted to humans.

Apart from the populations of farm animals that we have traditionally called domestic for many centuries, new species have begun to move into this category over recent decades: for example, some species of fish such as rainbow trout (Fauconneau, 2004). Domestication has come about very slowly, often in the countries of origin of the species, where the original wild species may have now disappeared. These animals then accompanied humans in their migration to other countries. In fact, very few populations of domestic animals in Western Europe originate from species that lived outside an association with humans: the pig, the rabbit, the dog, the grey goose and the pigeon are all examples of this. Most of Europe's farm species come from the Middle East (sheep, cattle, goats and cats), Africa (guinea fowl and Muscovy or Barbary duck), India (poultry), North America (turkey), South America (guinea pig) and East Asia (dog).

Domestication results from conscious or unconscious selection by breeders of animals that can accept the presence of humans and can be manipulated by them. One well-documented example of selection for docility and attraction to humans is the silver fox in the Soviet Union which was normally bred for its fur (Belyaev *et al.*, 1985). They found that some foxes were less fierce than others when approached by humans, and systematically interbred the least fierce foxes over 40 generations (Belyaev, 1969; Trut, 1999). This resulted in foxes that reacted positively to the approach of humans, and also waved their tails, headed towards them and sought contact. In fact, it is speculated that their experiment closely resembles, over a very much shorter period, the domestication of the dog and the emergence of the domestic dog (*Canis lupus familiaris*) from the grey wolf (*Canis lupus lupus*). One possible scenario for this speciation is that grey wolves may have approached human encampments in the distant past and somehow begun to domesticate themselves (Morey, 1994; Coppinger and Coppinger, 2002). That first phase which then led to wolves and humans getting closer together would have been followed by an artificial selection by humans to finalize the transition.

Belyaev's hypothesis underlying his experiment with Arctic foxes (another name for silver foxes) was that the formation of the dog species came from genetic change that modified its behaviour, a modification he successfully replicated in foxes. Belyaev (1969) also showed that the natural or artificial selection of traits like docility led to the changes in reproductive physiology and morphology that Darwin (1859) had emphasized earlier. Another such selection for behaviour was conducted in gerbils by Price (1984), who, in a few generations, produced animals that were able to live and reproduce under caged conditions. And, it has recently been shown that sheep can be selected successfully for calmness or for stressfulness (Qiu *et al.*, 2017).

Studies in cattle exemplify how domestic animals can vary according to the objectives of breeders (Le Neindre, 1989a, b). Environment and breed are important in determining how animals respond to humans (Boivin *et al.*, 1994). They show desirable behaviour towards humans only if their experiences, especially early ones, have been favourable for them. Cattle that have had early and positive contact with humans react less negatively to subsequent contact than those that had not. The neonatal period and weaning are two key periods for the establishment of this favourable reactive behaviour (Boivin *et al.*, 1992).

However, the definitions of domestication and taming are not clear-cut. Bees (*Apis mellifera*) may be a domestic species, but strictly speaking they are neither housed close to humans nor reproduced under total human control, except that beekeepers nowadays select queens. They are simply captured and housed in hives that are suitable both for bees to accommodate their larvae and food resources, and for humans to collect honey. House sparrows (*Passer domesticus*) that live close to human dwellings and benefit from human food remains are also called domestic. But species such as the Turkish turtle dove and herring gulls, which nest in the roofs of buildings and eat household food scraps, or even black rats, should also be described as domestic. But, in all these species, commensal with humans, their reproductive success is outside of human control. On the contrary, human activity towards them is often aimed at limiting or even eradicating their entire population.

Apart from cats raised in catteries (*Felis catus*) that would meet the criteria of being a domestic species, most cats reproduce without interference from humans and enjoy great freedom which is considered the basis of their well-being. Even though they are close to humans, cats are often considered more of a tame species than a domestic one.

The genetic modifications that humans seek to impose on domestic animals depend on the use they plan to make of them. Coppinger and Coppinger (2002) defined domestic as 'genetically close to humans' which included species that were either prey or more or less direct competitors such as wolves or ferrets. Keeping these species in confinement then allowed a phase of artificial selection oriented towards some desired morphological and/or behavioural trait. During this process behavioural traits such as docility or attractiveness to humans were always sought explicitly. For example, in cattle, good docility, or little or no adverse reaction to handling, is habitually sought given the large size of the animals and the potential danger in handling them if they are not docile. For dairy cows, high milk production is sought at the expense of maternal behaviour. In horses, artificial selection is oriented towards animals that are spirited but which remain docile and therefore not dangerous for those who ride them. This has become the essential role of the horse, a transport animal, mounted or harnessed. In pigs, the main characteristics sought are prolificity and size associated with conformation, although some breeds remain genetically close to the wild species (*Sus scrofa*). None the less, despite this intense selection, domestic pigs are still the same species, *S. scrofa*. Even so, in some species hypertypes have been developed in which the animals can now only live and reproduce under human control. For example, some dairy cows can now produce more than 10,000 l of milk per lactation, well beyond the volume needed to feed one or two calves. This has been achieved to the detriment of the viability of animals that now require specialized management and nutrient-enhanced diets. By contrast, most beef cows have simpler rations of fresh or preserved fodder although there are genetic variants like the so-called double-muscled breeds, Belgian Blue and Charolais, that produce extra meat (Vissac *et al.*, 1974). But this desirable trait is accompanied by difficult parturition and the need to assist most females at calving. In sheep, there is a similar mutation, the callipyge gene which produces large hindquarters, but leads to difficult births so it is rarely selected for. In pigs, particularly in the Piétrain strain, there are lines that grow well but are highly sensitive to stress. In poultry, as we have already seen, modern breeds of laying hens lay copiously but are no longer able to brood and rear chickens. So, for most species of farm animals, there are characters that would be lethal to them without the intervention of humans.

Dogs are a special case. We have seen over time a profound differentiation between them and wolves to create a new species, or at least a new subspecies because wolves and dogs can still intermate and produce fertile offspring (Ollivier, 2017). Humans have created dogs for a wide variety of uses by exploiting their wide diversity of behavioural potentialities, while taking advantage of their attraction towards humans who, in turn, encourage and develop it. These uses have been matched by a wide variety of morphologies including sizes and weights. Human imagination has even led to hypertypes where certain traits have been pushed to the extremes such as dwarfism, gigantism, and distortion of facial mass. This unbridled selection leads, or may potentially lead, to a form of eugenics in which individuals who do not conform strictly to arbitrary criteria set by humans, are systematically eliminated.

The cat, on the other hand, has been largely spared. The ecological utility of the cat in the fight against mice has not led humans to attempt to change this species greatly from the original *Felis lybica*. The only selection has been cosmetic for traits considered by humans as aesthetic, such as the face of Persian cats, the length and number of hairs, and a few other modifications to meet sometimes surprising cultural criteria. For example, breeds of tailless cats, like the Manx cat, carry a gene that is lethal in the homozygous state and causes disturbances in the heterozygous state (Buckingham *et al.*, 2013).

Nowadays, domestication and the confinement that it often implies, is a significant challenge for animal welfare in the face of significant changes in feeding patterns, parental care and possible activities for individuals. Selection for lack of apprehension towards humans probably focused on fundamental traits. Thus, for the dog, it is claimed that domestication was characterized by selection for neoteny, or the retention of juvenile features in the adult animal, that facilitates attachment with humans (Lorenz, 1957). Puppies, and maybe other animals, that have contact with humans from when they open their eyes soon after birth, subsequently make easier contact (Scott and Fuller, 1965). Recently, it has been argued that domestication of poultry was facilitated by an epigenetic modification of their reaction towards humans (Bélteky *et al.*, 2018).

The taming of individuals and the ease of taming animals are therefore powerful mechanisms that explain how domestic animals have gradually been able to thrive by accepting contact with humans. We are now beginning to understand the physiological and genetic mechanisms that explain these changes. Selection that focuses on production traits has often had a severe impact on the animals' behaviour and their welfare. It is therefore misleading to look only at the natural habitat of the ancestors of domestic species for defining ways of how animals should adapt to the constraints of the artificial environments in which they currently live.

5.3   Conclusion: the Way Forward

So far, our review of the biological knowledge underpinning animal behaviour is predicated on two important precepts. The first is that we must define the capacity of animals to perceive their world; that perception is highly variable, and their capacity is often far greater than that of humans. The second is that it is possible to understand better the subjective world of animals even if some authors think that we will never be able to measure objectively and scientifically 'the subjective and private nature of conscious experience' (Dawkins, 2014). Dawkins claims that it is better to leave the question of animal consciousness to the academics and to 'study conditions that improve the animal health, their immune systems, their disease resistance, and other conditions that have positive impact on human well-being' (p. 27). Milhaud (2007) introduced the idea of 'well-treating', or *bien-traitance* in French, which is to follow our own ideas for the best way to act on behalf of animals. However, thanks to new experimental results, we can now base more of our analysis on the idea that animals have a consciousness (Broom, 2014; Le Neindre *et al.*, 2017). What we described in the previous chapter is that they perceive emotions, pain, suffering and pleasure in real time and they have expectations for their future (Le Neindre *et al.*, 2009). Animals are not only passive actors that can tell us what they experience but actors that can tell us about their life. Broom (1996) defined animal welfare in terms of their behavioural attempts to cope with their environment and concludes by linking that behaviour with human activity.

The published literature so far concerns only a few animals in a few situations, but it is still sufficient to allow us to extrapolate to broader contexts. Some authors think that we still do not have a sound rationale for using them to shape our action (Dawkins, 2014). Even so, we think it is a starting point for considering what we already know to develop sound regulatory, societal and economic decisions in our society. The lack of knowledge should not be an excuse for inaction as described in another context by Oreskes and Conway (2015).

As a result, we suggest that scientists have a new role in the debate. 'Experts' need to clarify society's concepts of what is at stake and their strengths and weaknesses (Roqueplo, 1997). Scientists may not be involved in the political decisions, but they should be aware of how their writings, especially their conclusions, are used. They may not be the judges, but they are the auxiliaries of the judges. Some scientists think that they should go further and be involved in decisions of society, and that they should be advocates for the animals. That trend is very active in the USA where several animal case studies have been published (Carrié and Traïni, 2019).

References

Bélteky, J., Agnvall, B., Bektic, L., Höglund, A., Jensen, P. *et al.* (2018) Epigenetics and early domestication: differences in hypothalamic DNA methylation between red junglefowl divergently selected for high or low fear of humans. *Genetics, Selection, Evolution* 50(1), 13. DOI: 10.1186/s12711-018-0384-z.

Belyaev, D.K. (1969) Domestication of animals. Science 5, 47–52.

Belyaev, D.K., Plyusnina, I.Z. and Trut, L.N. (1985) Domestication in the silver fox (*Vulpes fulvus* Desm): changes in physiological boundaries of the sensitive period of primary socialization. *Applied Animal Behaviour Science* 13(4), 359–370. DOI: 10.1016/0168-1591(85)90015-2.

Boivin, X., Neindre, P.L. and Chupin, J.M. (1992) Establishment of cattle-human relationships. *Applied Animal Behaviour Science* 32(4), 325–335. DOI: 10.1016/S0168-1591(05)80025-5.

Boivin, X., Le Neindre, P., Garel, J.P. and Chupin, J.M. (1994) Influence of breed and rearing management on cattle reactions during human handling. *Applied Animal Behaviour Science* 39(2), 115–122. DOI: 10.1016/0168-1591(94)90131-7.

Bonta, M., Gosford, R., Eussen, D., Ferguson, N., Loveless, E. *et al.* (2017) Intentional fire-spreading by "firehawk" raptors in Northern Australia. *Journal of Ethnobiology* 37(4), 700–718. DOI: 10.2993/0278-0771-37.4.700.

Boyer, A. and Planiol, M. (1948) *Traité de Fauconnerie et Autourserie. Avec 33 Dessins Originaux de R. Reboussin, 13 Illus.* Payot, Paris.

Broom, D. (1996) Animal welfare defined in terms of attempt to cope with the environment. *Acta Agriculuræ Scandinavica, Section A, Animal Science Supplementum* 27, 22–28.

Broom, D.M. (2014) *Sentience and Animal Welfare.* CAB International, Wallingford. DOI: 10.1079/9781780644035.0000.

Buckingham, K.J., McMillin, M.J., Brassil, M.M., Shively, K.M., Magnaye, K.M. *et al.* (2013) Multiple mutant T alleles cause haploinsufficiency of Brachyury and short tails in Manx cats. *Mammalian Genome* 24(9–10), 400–408. DOI: 10.1007/s00335-013-9471-1.

Carrié, F. and Traïni, C. (2019) *S'engager Pour les Animaux.* Presses universitaires de France (PUF), Paris, 116 pp.

Clutton-Brock, J. (1999) *A Natural History of Domesticated Mammals,* 2nd edn. Cambridge University Press, Cambridge, 238 pp.

Coppinger, R. and Coppinger, L. (2002) *Dogs: A Startling New Understanding of Canine Origin, Behavior and Evolution.* University of Chicago Press, Chicago, Illinois.

Darwin, C. (1859) *On the Origin of Species by Means of Natural Selection, or, The Preservation of Favoured Races in the Struggle for Life.* John Murray, London. DOI: 10.5962/bhl.title.82303.

Dawkins, M. (2014) Animal welfare and the paradox of animal consciousness. Advances in the Study of Behavior 47, 5–38.

Denis, B. (coord.) (2003) Animal domestique, espèce domestique, domestication: points de vue. *Ethnozootechnie* 71, 154.

Denis, B. (2007) *Génétique et Sélection Chez le Chien.* Pratique Médicale et Chirurgicale de l'Animal de Compagnie (PMCAC), Paris, 320 pp.

Deputte, B.L. and Busnel, M. (1997) An example of a monkey assistance program: P.A.S.T. – The French Project of Simian Help to Quadriplegics. *Anthrozoös* 10(2–3), 76–81. DOI: 10.2752/089279397787001201.

Digard, J.P. (1988) Jalons pour une anthropologie de la domestication animale. *L'Homme* 28(108), 27–58. DOI: 10.3406/hom.1988.369042.

Digard, J.-P. (1992) Un aspect méconnu de l'histoire de l'Amérique: la domestication des animaux. *L'Homme* 122–124, 253–270.

Farm Animal Welfare Council (1979) Farm Animal Welfare Council. Press statement, 5 December. Available at: http://webarchive.nationalarchives.gov.uk/2 0121007104210/http:/www.fawc.org.uk/pdf/fivefreedoms1979.pdf (accessed 10 May 2022).

Fauconneau, B. (2004) Diversification, domestication et qualité des produits aquacoles. *INRAE Productions Animales* 17(3), 227–236. DOI: 10.20870/productions-animales.2004.17.3.3596.

Faure, J.M. and Le Neindre, P. (2009) Domestication des espèces animals. In: Boissy, A., Pham-Delègue, M.-H. and Baudoin, C. (eds) *Éthologie Appliquée: Comportements Animaux et Humains, Questions de Société.* Quae, Versailles, France, pp. 56–66.

Frantz, L.A.F., Mullin, V.E., Pionnier-Capitan, M., Lebrasseur, O., Ollivier, M. *et al.* (2016) Genomic and archaeological evidence suggest a dual origin of domestic dogs. *Science* 352(6290), 1228–1231. DOI: 10.1126/science. aaf3161.

Geoffroy Saint-Hilaire, I. (1861) *Acclimatation et Domestication des Animaux Utiles.* La Maison Rustique, Paris, 158 pp.

Hansen, H.R., Hector, B.L. and Feldmann, J. (2003) A qualitative and quantitative evaluation of the seaweed diet of North Ronaldsay sheep. *Animal Feed Science and Technology* 105(1–4), 21–28. DOI: 10.1016/S0377-8401(03)00053-1.

Hemmer, H. (1990) *Domestication: The Decline of Environmental Appreciation.* Cambridge University Press, Cambridge, 208 pp.

Labussière, J. and Richard, P.H. (1966) La traite mécanique. Aspects anatomiques, physiologiques et technologiques. Mise au point bibliographiques. *Annales de Zootechnie, INRA/EDP Sciences 1965* 14, 63–126.

Le Gros, M. (2017) *Eloge de La Palourde, La Nouvelle Escampette.* Librairie les saisons, La Rochelle, France, p. 139.

Le Neindre, P. (1989a) Influence of cattle rearing conditions and breed on social relationships of mother and young. *Applied Animal Behaviour Science* 23(1–2), 117–127. DOI: 10.1016/0168-1591(89)90012-9.

Le Neindre, P. (1989b) Influence of rearing conditions and breed on social behaviour and activity of cattle in novel environments. *Applied Animal Behaviour Science* 23(1), 129–140. DOI: 10.1016/0168-1591(89)90013-0.

Le Neindre, P., Bernard, E., Boissy, A., Boivin, X., Calandreau, L. *et al.* (2017) *Animal Consciousness*. EFSA (European Food Safety Authority) supporting publication, Parma, Italy, 165 pp.

Le Neindre, P., Guatteo, R., Guémené, D., Guichet, J.L., Latouche, K. *et al.* (2009) *Douleurs Animales. Les Identifier, Les Comprendre, Les Limiter Chez Les Animaux d'élevage. Expertise Scientifique Collective, Rapport d'expertise.* INRA, Paris, 340 pp.

Lésel, R. (1969) Etude d'un troupeau de bovins sauvages vivant sur l'île d'Amsterdam. *Revue d'Elevage et de Médecine Vétérinaire Des Pays Tropicaux* 22, 107–125.

Leslie, D.M. and Schaller, G.B. (2009) *Bos grunniens* and *Bos mutus* (Artiodactyla: Bovidae). *Mammalian Species* 836, 1–17. DOI: 10.1644/836.1.

Lorenz, K. (1957) *Essais sur le Comportement Animal et Humain*. Seuil, Paris, 477 pp.

Milhaud, C. (2007) Rapport sur l'utilisation du néologisme « *bientraitance* » à propos de la protection des animaux. Available at: http://academieveterinaire.free.fr /rapports/bientraitance.pdf (accessed 10 May 2022).

Mills, A.D., Jones, R.B. and Faure, J.M. (1995) Species specificity of social reinstatement in Japanese quail *Coturnix japonica* genetically selected for high or low levels of social reinstatement behaviour. *Behavioural Processes* 34(1), 13–22. DOI: 10.1016/0376-6357(94)00044-h.

Morey, D. (1994) The early evolution of the domestic dog. *American Scientist* 82, 336–347.

Ollivier, M. (2017) Reconstruire et comprendre l'histoire de la domestication du chien grâce à la paléogénétique. *Archéozoologies* 148, 50–55.

Oreskes, N. and Conway, E. (2015) *Merchants of Doubt: How a Handful of Scientists Obscured the Truth on Issues from Tobacco Smoke to Global Warming*. Bloomsbury Press, London, 400 pp.

Price, E.O. (1984) Behavioral aspects of animal domestication. *The Quarterly Review of Biology* 59(1), 1–32. DOI: 10.1086/413673.

Price, E.O. (1999) Behavioral development in animals undergoing domestication. *Applied Animal Behaviour Science* 65(3), 245–271. DOI: 10.1016/S0168-1591(99)00087-8.

Qiu, X., Martin, G.B. and Blache, D. (2017) Gene polymorphisms associated with temperament. *Journal of Neurogenetics* 31(1–2), 1–16. DOI: 10.1080/01677063.2017.1324857.

Roqueplo, P. (1997) *Entre Savoir et Décision, l'Expertise Scientifique*. Sciences en questions, Editions Quæ, Versaille, France, 112 pp.

Sánchez Chillon, B., Prado, J.L. and Alberdi, M.T. (2003) Paleodiet, ecology, and extinction of pleistocene gomphotheres (proboscidea) from the pampean region (argentina). *Coloquios de Paleontología* 1, 617–625.

Sauveur, B. (1988) *Reproduction des Volailles et Production d'œufs*. INRA, Paris, 472 pp.

Scott, J.P. and Fuller, J.L. (1965) *Genetics and the Social Behavior of the Dog*. University of Chicago Press, Chicago, Illinois.

Trut, L. (1999) Early canid domestication. *American Scientist* 87, 160–169.

Vernon, P. (1986) Evolution des réserves lipidiques en fonction de l'état physiologique des adultes dans une population expérimentale d'un diptère subantarctique, anatalanta aptera eaton (sphaeroceridae). *Bulletin of the Society of Ecophysiology* 11(1), 95–116.

Vernon, P. and Vannier, G. (1996) Developmental patterns of supercooling capacity in a subantarctic wingless fly. *Experientia* 52(2), 155–158. DOI: 10.1007/BF01923362.

Vissac, B., Perreau, B., Mauléon, P., Ménissier, F., de Fontaubert, Y. *et al.* (1974) Etude du caractère culard IX. Fertilité des femelles et aptitude maternelle. *Annales de Génétique et de Sélection Animale, INRA Editions* 6, 35–48.

Wayne, R.K. and Ostrander, E.A. (2007) Lessons learned from the dog genome. *Trends in Genetics* 23(11), 557–567. DOI: 10.1016/j.tig.2007.08.013.

Woinarski, J., Legge, S. and Dickman, C. (2019) *Cats in Australia 2. Companion and Killer.* CSIRO Publishing, Clayton, Victoria. DOI: 10.1071/9781486308446.

Animal Welfare: the Responsibility of Humans for Animals

6

The reaction of the animal to its environment has two components: (i) sensitivity; and (ii) well-being. According to the *Grand Larousse encyclopédique* (Augé, 1964, p. 750), 'sensitivity is the property of a living organism to experience physical impressions'. It can be simply a reflex, and independent of any mental state as if we were talking about the sensitivity of a machine or a measuring tool. There are even sensitive plants such as *Mimosa pudica* that reacts to touch, and so meet the criterion of being sensitive. The well-being of the animal is more complex and requires it to make mental analyses that allow it to answer a series of questions. The first question is whether the situation deserves attention at all. The second is whether, because of the experience of the animal or because of its species, it will produce a positive or negative response. This response determines the way the animal senses the situation and finally reacts in the way it wishes.

Animals can tell us their level of well-being, not through language but through their activities, health, morbidity and reproduction. So we can question animals by observing their behaviour and physiology, providing that the constraints we put on them to do so don't harm their interest or what they would normally experience. We can compare this sort of questioning to Canguilhem's definition of human health: 'life in the silence of organs' (Canguilhem, 1943, p. 158). In other words, he is saying that only when no damage is found in the organs one is investigating is it possible to make feasible diagnoses. Similarly, animals need to be in circumstances as normal as possible before we can conclude that their animal welfare is not threatened, but even then, we can't be entirely sure.

The French national Agency for Food, Environmental and Occupational Health and Safety (ANSES) proposed a definition of well-being based on recent scientific information about cognitive ethology (Mormede *et al.*, 2018). According to this definition, 'the well-being of an animal is its positive mental and physical state when it is satisfied with its physiological and behavioural needs and expectations'. This state varies according to the animal's perception of what is happening around

© CAB International 2022. *Interacting with Animals: Understanding their Behaviour and Welfare* (Pierre Le Neindre and Bertrand L. Deputte) DOI: 10.1079/9781800622418.0006

it (Mormede *et al.*, 2018, p. 154). This definition considers both the negative and the positive aspects associated with pleasure, as opposed to displeasure, and the expectations of the animal. It also takes account of the animal's expectations which has been demonstrated for example on the grunts of pigs (Villain *et al.*, 2020). Although researchers have sought for a long time to minimize suffering in animals, Yeates and Main (2008) and more recently Baciadonna *et al.* (2018) highlighted the importance of maximizing pleasures, minimizing displeasure, and taking into account the animal's emotions. Balcombe (2009) also emphasized the need to draw moral consequences when observing behaviour.

Another way of analysing animal behaviour is by using an operational grid. This grid sets out five freedoms (FAWC, 1979) and these are now included in the World Organisation for Animal Health (OIE) definition of animal welfare. The five freedoms are: (i) freedom from hunger and thirst; (ii) freedom from discomfort (e.g. physical and thermal stress); (iii) freedom from pain, injury and disease; (iv) freedom from fear and distress; and (v) freedom to express the normal behaviours of their species. This last freedom is the only one expressed positively. It captures the notion of being natural or behaving in a way that is important for animals because of their evolutionary history. Achieving this last freedom is the most difficult operationally because it involves knowing the elements of the environment that animals find important.

The list in the grid is valuable because it differentiates potential avenues for making progress. We can now target possible points on which to concentrate our attention, but it still does not alert us to thresholds beyond which we should be concerned. In other words, we can say that an animal should not be hungry, but we can't define the limit below which low consumption induces hunger within the continuum between great hunger and absolute satiety. In the end, alert zones can only be portrayed by observing how animals react. In addition, these alert zones may vary depending on the animal's age and physiological stage of development. For example, hunger in a brooding hen cannot be defined in the same way as in a hen that is producing eggs. And pain is not the same for a hen that is about to lay an egg and for another that is at rest.

The relative importance of each of these items is assessed by studying physiological or behavioural consequences when they are absent. A symbolic example of this type of research is the analysis of stereotypies (repetitive movements). This is an aberrant behaviour that is often seen when farming systems impose constraints on animals. For example, veal calves in very restrictive and poor environments sometimes develop stereotypies such as unusual tongue movements and also gastric ulcers (Wiepkema *et al.*, 1987) which are both indicators of psychological suffering. Suffering is defined as the emotional state of distress associated with events that threaten the physical or psychological integrity of the individual (Le Neindre *et al.*, 2009). Another definition is: 'animal pain is an aversive sensory and emotional experience representing an awareness by the animal of damage or threat to the integrity of its tissues' (Molony and Kent, 1997, p. 266). These definitions include not only the physical pain that the animal feels but also psychological suffering, such as separation from its mother. At the human level, this type of psychological suffering is clearly illustrated if a mother is separated from her child.

None the less, there are many clues that may help us understand how animals interpret the situation they are experiencing. Animals can vary the rhythms of their activity depending on the environments in which they are placed, and these are best studied in a comparative analysis. This is illustrated by an experiment in which cows were housed in buildings with contrasting floor surfaces such as concrete, grating or straw. Behaviour was observed by collecting data about the time cows spent lying down and data relating to health including injuries and inflammation of the limbs. Variations in these criteria reflected the relative discomfort associated with other aspects, like the stalls being too small or the type of flooring that might injure them (Cazin *et al.*, 2014).

We can also analyse what happens to animals' behaviour when we change their environment. We can give housed cattle access to new surroundings such as to an outdoor area, possibly a pasture, or give chickens a new nest. New objects can be introduced such as scratching pans for chickens, or objects, often called 'toys', that attract attention and stimulate animal activity (Meunier-Salaün, 2017). The impact of these innovations is all the more important when the original environment offers little stimulation because a new object in an already rich environment arouses little interest, and then only for a short time. Animals may also have to work to access resources and this technique is often used in behavioural research. In the case of hens, this work may consist of pushing or pecking buttons. Cows can also show how they choose alternative feedstuffs, hens how they choose space, minks how they find access to water, and sheep how they choose social partners. The analysis of behaviour and its disorders, however, always requires caution to verify that the observer's interpretation is something like that of the animals they are observing. This interpretation must always consider the species of animal and be formed in the context of the animal's past history.

Of course, there are some pitfalls to be avoided. Thus, tail movements are a sign of contentment in dogs but dissatisfaction in cats. Similarly, a lack of behavioural response can give vital clues. Some animals, when they suffer, cease their activities altogether, and sometimes hide, while others will whine or become aggressive. Thus, cows that have multiple problems with their legs stop walking, which may be translated by the observer as the absence of a problem even though the cows may be suffering significantly. A cow that has one painful leg will limp but when she has two or more painful legs she no longer limps, but that does not mean that she is no longer in pain.

Animals, such as mice that may be prey to other animals, become immobile and feign death in the face of danger (Mills *et al.*, 1995). This tonic immobility is, in fact, an absence of reaction and is sometimes accompanied by relaxed muscles. But it does not signify that the animal is unaware of its situation but that it has acted with an almost reflex response that allows it, with luck when the predator's attention turns away, to escape.

All experiments into animal welfare must ensure that animals have the information and physiological resources to answer the questions we ask of them. We saw earlier that some animals may not have the capacity to perceive the elements that they are confronted with and so do not perceive the same images, sounds and smells as we do. On the other hand, many can perceive noises or smells that we humans cannot perceive, and that can lead to our drawing the wrong conclusions from the information we gain if we are not careful.

Physiological characteristics may not allow researchers to determine the right signals they are looking for. An example is that of pain related to force-feeding to produce *foie gras*. Ducks and geese are forced to receive food by intubating their digestive tract so that they ingest much more food than they would have ingested spontaneously. Muscovy and mule ducks, and grey Landes and Toulouse geese, store fat in their liver and not in the skin like most other birds. This is done by increasing the size of and not the number of hepatocytes, or liver cells. This state does not correspond to what is called steatosis or fatty liver in humans. However, when pathologists in human medicine examined the livers of force-fed ducks, they concluded that the same livers in humans would be considered to come from individuals with pathology. By contrast, in ducks the cells do not die and, in fact, the increase in the size of hepatocytes is reversible. This means that linking the state of liver tissues to pain is difficult because the liver does not have nociceptors through which the birds would perceive suffering in their liver tissues (Le Neindre *et al.*, 1998). It does not mean that other features of those production systems, especially the intensive ones, are not problematic from a welfare point of view.

Behaviour is the best tool for measuring the living conditions of animals. It makes it possible to understand how animals feel about their living environment. We can use ethology, combined with other tools like their physiology and neurophysiology, to give us clues about what they want to avoid and what they are seeking to do.

6.1   Animal Welfare and Human Involvement

One of the difficulties in trying to combine all these measures is that they do not all follow the same logic. Non-scientists are often out of their depth when trying to follow scientific discussions that concentrate on flaws in the arguments that lead to conclusions, rather than the coherence of the arguments. In fact, in science, the rule is to highlight doubts about the concepts, methods and conclusions of other scientists. Only when all contradictory arguments have been reviewed and refuted can acceptable conclusions be drawn. And, even then, these conclusions are still tentative while waiting for new arguments to shake them up. To handle this dilemma, Milhaud (2007) has suggested replacing the word, *well-being*, with '*bientraitance*' or *good treatment*. Dawkins (2014) proposed the same idea. These two meanings, one describing state and the other action, are integrated in the English word, *welfare*. Milhaud doubts that it is possible to evaluate the mental state of animals, or that it is of much interest to evaluate them, but above all highlights the influence of humans on the living conditions of animals. We don't believe that this is appropriate and maintain that we need to assess the state of comfort of the animal if we are to be effective in defining the areas that can be improved. An effective action, for a good treatment, requires a sound welfare evaluation.

The definition of *welfare* proposed by Mormede *et al.* (2018) seems to us to be a complete paradigm shift. It prompts us to change from choosing deductively the best solutions, without having hypotheses, to thinking inductively, by testing hypothesis (Marty, 1998). The two approaches are both rational but meet different objectives. The choice of which one to use cannot be done realistically without the proponents defining their real objectives, in terms of what they know and what they want to do.

It seems to us that the controversy over animal welfare is largely generated by the lack of clarity about these two approaches. Thus, some have said that well-being is a bad concept because it cannot be defined irrefutably and definitively on a quantitative basis. This is one of the arguments used to advocate *good treatment* as a term which, clearly, does not concentrate on the animal but on the human, in particular the breeder. We should look at all arguments, but there will never be a standard to distinguish good from bad – terms that carry moral dimensions that are outside the field of science. Nevertheless, these arguments are no justification for neglecting the behaviour of animals.

Animal welfare must be analysed with discernment to avoid both anthropomorphism, or projecting our own views onto that of other animals, and anthropocentrism, or reflecting our personal wishes and wills. It has legislative, regulatory and legal components, all of which are important, and which use concepts from biology and ethology without really explaining the scientific elements underlying them. Therefore, we have to know how to distinguish scientific elements from the others which are equally legitimate, but which are not subject to the same logic.

6.2   Who Acts For and On Behalf of Animals?

Many people are or would like to be involved in the debate about who acts for and on behalf of animals, and they formulate their own wishes and sanctions. A difficulty is that each stakeholder can have several motivations simultaneously that are often not made clear. We can separate these stakeholders into two categories. In the first group are those with an economic interest. It includes breeders, veterinarians, traders and distributors. The second is intended to be our moral conscience, endeavouring to inject the concept of good and evil, by seeking better living conditions for animals, ranging from as far as the extreme view of vegans who advocate that there should be no farm animals at all to that of contemporary farming and supply-chain practices.

6.2.1   Breeders

Every day, breeders make decisions that affect the quality of life of animals. These decisions have an economic component, but breeders also need to consider the physiology and health of the animals for which they are responsible. However, these decisions also depend on the interactions that breeders themselves may have with animals. Interactions between breeders and animals are often emotional. This emotional connection is frequently unexplained and can even be hidden. They are even more hidden because they go against simple economic rationality. This emotional concern of breeders is seen by some as the essential and profound characteristic that makes a producer a breeder (Porcher, 2014). Breeders will openly say that they are keeping an animal that one would expect to be culled on simple technical or economic criteria. 'I'm going to give her a year to recover because I like her' or 'I'm hesitant to cull her because she reminds me of the birth of my eldest child.' So breeders decide on a wide range of actions that affect animals and try to justify their actions (for example Laporte and Mainsant, 2012). They decide on which animals should be bred, kept or culled and the design of their livestock buildings and milking parlours that will determine for years the quality of life of the animals and their health. This includes the flooring, the space per animal, the design of laneways and especially food. They also decide whether or not to castrate young males knowing the short- and long-term consequences for them. In short, they are 'the missing link' between the animals and society (Grandin, 2014).

6.2.2   Veterinarians

Breeders base many of their decisions on advisors, particularly veterinarians who give guidance and opinion on matters of health and ensuring well-being (Thiermann and Babcock, 2005). They and the breeders are usually the only ones to have physical contact with animals to bring into play practices that are sometimes fraught with serious consequences. They understand, to some extent, the relationship of the breeders with their animals and can say of a breeder that his or her animals are in difficulty because he or she has personal problems. However, understanding the breeder takes time and not only relies on the empathic links between breeders and animals, but on the limited time that the veterinarian can spend on the farm.

6.2.3   Banks and agribusiness interests

Other actors in the technical and financial spheres associated with farming can sway the choices that breeders make. The banker influences the type of equipment and housing that the breeder can afford. Technical advisors, who are often allied with private companies or professional cooperatives, also have an impact on breeders' decisions. Generally, these people have no emotional concern at all for the animals and their contributions are based only on financial rationality. Despite this influence of bankers and technical advisors, commercial factors concerning the marketing and sale of the farm produce are increasingly becoming a factor. More and more animal-based products are offered for sale with captions that highlight the well-being of the animals from which they come. For example, some brands of milk are described as 'ethical'. Hens and chickens are raised in 'free-range conditions' and on farms that 'respect animal welfare'. The specifications of products labelled 'organic' often include reference to a component of well-being. Distributors are currently partnering with animal welfare societies to offer 'welfare' labels and brands. Guides to good practice are published that demonstrate that professionals are involved to avoid distressing situations on-farm, during transport and in slaughter-houses. The government does not initiate these labels, but it must assure consumers that the information in these publications is genuine.

Distributors and manufacturers are also increasingly involved in the future of livestock sectors. In response to consumers' views, they show their willingness to improve and promote new breeding practices by including criteria relating to this concern. This has resulted in promotion of animal welfare in 'organic' farming or 'biodynamic' farming, or by offering special products such as uncastrated male pigs or non-debeaked and non-caged laying hens. The impact of this is likely to be increasingly influential in the future (Botreau and Veissier, 2012; Sørensen and Schrader, 2019).

6.2.4   Expert and advisory bodies

Specifications put forward by many of the stakeholders above are based on qualifications that are sometimes poorly defined and this can leave consumers perplexed and often doubtful as to the soundness of the arguments being put forward. Researchers, in particular, are asked by the

European Commission to be more active in explaining their work and the way that they arrive at their conclusions. Public authorities, too, are trying to do more to ensure that their specifications are transparent and easy to understand.

In view of this network of restrictions and claims, some researchers, including us, the authors of this book, believe that researchers should be independent of stakeholders and be able to analyse the living conditions of animals throughout their complete life cycle. In any case, suggestions researchers make should be submitted in a transparent manner for the scrutiny of all stakeholders. Institutions responsible for analysing the future of livestock farming, such as agricultural and veterinary academies, often advise decision-making bodies without explaining their links to stakeholders, in particular with professionals. Professional researchers in biology, like us, also have a perspective on animals. Their interests are often purely academic, but like everyone else, they may have parallel motives, even if these are only prestige or promotion.

6.2.5 Animal protectors

At the social level, societies for the protection of animals, or welfarists, appeal to the public to advocate suppression of certain practices or promotion of others. Sometimes, what they advocate for the way we behave with animals is extreme. They question so-called 'industrial' farming which in their eyes mistreats animals as well as the people who look after them. Their criticisms involve all aspects of the sector – genetics, selection, housing and feeding, transport and slaughter. They denounce practices that they deem intolerable by informing the public and political decision makers through press campaigns or through social networks. Nevertheless, they sometimes use scientific knowledge on which to base their analyses (Baussier, 2021) and without doubt, may be important and positive drivers to stimulate effective change. On the other hand, other groups are trying to radically change our entire relationship with animals. Among them, vegans advocate that all uses of animals, whether for food, pleasure or work, should cease. This would mean abolishing all farming involving the raising of animals and all intervention in the delicate management of those animals that don't directly depend on humans.

6.2.6 Legal and political bodies

In the last 30 years, the influence of jurisdictions on the welfare of animals has increased greatly. To illustrate this, let us look at some of the changes in how the French judicial system regards animals.

Animals were once described simply by their physical characteristics and declared as goods that belonged to a person or group of people. Nowadays, regulations in the French Rural Code specifically refer to them as 'animals', and Article L214 defines in legal terms that 'an animal is a sentient being that must be placed by its owner in conditions compatible with the biological requirements of its species' (French Government, 2015a, Article L214-1). In another section, L214-3, it states that the owner is criminally responsible if the animal suffers, and may incur fines and possibly imprisonment. It also uses biological terminology such as 'sentience' and 'biological imperatives of its species'. Article 515-14 of the French Civil Code from the French Government adopted in 2015 changed the rules governing the status of humans and the animals they might possess. That provoked a significant political reaction when it defined animals as 'living beings endowed with sensitivity' while still specifying that they are a 'property' (French Government, 2015b, Article 515-14). It was the first time that concern for animals had appeared in the French Civil Code. The Code also stated that 'the owner, or a person in charge of an animal, is responsible for any damage suffered by that animal regardless of whether it was in their custody, or was lost or had escaped' (Article 1385 of the French Civil Code). Some French politicians have been more active at the political level asking for further improvements in the production supply chain at the European level (Durand and Christophe, 2019).

At the European level, important changes have also taken place. Previously, the law was content to limit its deliberations to preventing distortions of competition between countries. Since then, the quality of life of animals has now become important. The law has also expressed its willingness to contemplate the effects of the production supply chain on animals. It considers not only animal products but also how they are produced. The Council of Europe has produced regulations and directives that have led to a formal structure. To support the decisions they make, administrations mobilize researchers to review what already exists at the scientific and social levels. Those analyses are done by the European Food Safety Authority (EFSA) for Europe, and ANSES for France.

This overview of the European and French systems demonstrates that animals are now the subject of much legal interest but their judicial status is still the subject of controversy (Desmoulin-Canselier, 2009). This is obvious in the drafting of most of the French and European laws and regulations. However, they promote some practices through financial incentives or through funding for research.

The European authorities have been particularly active in instituting laws and regulations that limit or prohibit certain practices. These relate to the housing of production animals, including laying hens, broilers, dairy cows, veal calves and pigs. They are also concerned with transport and slaughter.

In summary, many claims, and actions from different stakeholders are based upon the behavioural and cognitive abilities of the animals but they seem extremely unclear to a biologist. In particular, in the French Civil Code, the concept of 'being' is not defined. One could imagine that it involves not only animals, or some animals, but also plants (and why not robots?). Readers of these rules, which define animals as 'living beings endowed with sensitivity', would find it hard to distinguish whether only sentient beings are concerned or whether only animals are being described in this new way because they are intrinsically endowed with sensitivity.

Very often the rules state that we must avoid constraints and unnecessary suffering that we impose on animals, but it is not clear who defines what is necessary or unnecessary. From the animal's point of view, castration is probably unnecessary. From the consumer's point of view castration is considered by some as necessary for avoiding off-flavour in meat. The degree to which an action is necessary, or not, may be defined either as its impact on subsequent behaviour or by how it affects productivity. Clearly it is a concept that ought to be better defined and should be used sparingly.

Summing up, a lot of people contribute to how animals live. Breeders are probably the most important of these, but others contribute by influencing what the breeder does. Each has their own frame of reference and objectives, which may be ethical, economical, or both. We believe strongly that the best and most effective balance is always to refer to the animals themselves and what they are doing. Preference should be given to scientific evidence rather than pragmatic schemes that may be more efficient in the short term, but which do not define avenues of progress that are acceptable to all stakeholders.

References

Augé, C. (1964) *Grand Larousse Encyclopédique*, Vol. 9. Libraire Larousse, Paris, 750 pp.

Baciadonna, L., Duepjan, S., Briefer, E.F., Padilla de la Torre, M. and Nawroth, C. (2018) Looking on the bright side of livestock emotions – the potential of their transmission to promote positive welfare. *Frontiers in Veterinary Science* 5, 218. DOI: 10.3389/fvets.2018.00218.

Balcombe, J. (2009) Animal pleasure and its moral significance. *Applied Animal Behaviour Science* 118(3–4), 208–216. DOI: 10.1016/j.applanim.2009.02.012.

Baussier, M. (2021) *La Science Face à la Conscience … Animale: Évaluation et Amélioration du Bien-Être Animal*. Editions Book-e-book, Paris, 84 pp.

Botreau, R. and Veissier, I. (2012) Welfare Quality® – an innovative approach to animal welfare. In: *Presented at workshop 'Innovation in Livestock Farming System,' Institut National de la Recherché Agronomique (INRA), Paris*.

Canguilhem, G. (1943) *Essai sur Quelques Problèmes Concernant le Normal et le Pathologique, Réédité Sous le Titre le Normal et le Pathologique, Augmenté de Nouvelles Réflexions Concernant le Normal et le Pathologique (1966)*. 9ᵉ rééd. PUF/Quadrige, Paris.

Cazin, B., Nicks, B. and Dufrasne, I. (2014) Aménagement des logettes et confort des vaches laitières. *INRAE Productions Animales* 27(5), 359–368. DOI: 10.20870/ productions-animales.2014.27.5.3083.

Dawkins, M. (2014) Animal welfare and the paradox of animal consciousness. *Advances in the Study of Behavior* 47, 5–38.

Desmoulin-Canselier, S. (2009) Quel droit pour les animaux? Quel statut juridique pour l'animal? *Pouvoirs* 131(4), 43–56. DOI: 10.3917/pouv.131.0043.

Durand, P. and Christophe, M. (2019) *L'Europe des Animaux: Utiliser le Levier Européen pour la Cause Animale.* Alma, London, p. 227.

Farm Animal Welfare Council (1979) Farm Animal Welfare Council. Press statement, 5 December. Available at: http://webarchive.nationalarchives.gov.uk/2 0121007104210/http:/www.fawc.org.uk/pdf/fivefreedoms1979.pdf (accessed 10 May 2022).

French Government (2015a) Code rural et de la pêche maritime: Chapitre IV: La protection des animaux … (Articles L214-1 à L214-23). Légifrance. Available at: https://www.legifrance.gouv.fr/codes/id/LEGISCTA000006152208/ (accessed 1 June 2022).

French Government (2015b) Code civil: Livre II: Des biens et des différentes modifications de la propriété (Articles 515-14 à 710-1). Légifrance. Available at: https://www.legifrance.gouv.fr/codes/article_lc/LEGIARTI000030250342 / (accessed 1 June 2022).

Grandin, T. (2014) Animal welfare and society concerns finding the missing link. *Meat Science* 98(3), 461–469. DOI: 10.1016/j.meatsci.2014.05.011.

Laporte, R. and Mainsant, P. (2012) *La Viande Voit Rouge.* Librairie Arthème Fayard, Paris, 220 pp.

Le Neindre, P., Guatteo, R., Guémené, D., Guichet, J.L., Latouche, K. *et al.* (2009) *Douleurs Animales. Les Identifier, Les Comprendre, Les Limiter Chez Les Animaux d'élevage. Expertise Scientifique Collective, Rapport d'expertise.* INRA, Paris, 340 pp.

Le Neindre, P. (rapporteur), Willeberg, P., Jensen, P., Broom, D.M., Hartung, J. *et al.* (1998). Available at: https://ec.europa.eu/food/system/files/2020-12/s ci-com_scah_out17_en.pdf (accessed 23 June 2022).

Marty, F. (1998) Action économique et adaptations rationnelles: gestion par les firmaes agro-alimentaires d'un produit protégé soumis à un règlement technique. PhD thèse, Université Paris X, Nanterre, France.

Meunier-Salaün, M.-C. (2017) Impact d'un enrichissement du milieu de vie sur le bien-être chez le porc. In: *47. Colloque Annuel de la Société Française pour l'étude du Comportement Animal,* May 2017, Gif-Sur-Yvette, France. 2017, Sfeca 2017. Available at: https://hal.archives-ouvertes.fr/hal-01605852 (accessed 10 May 2022).

Milhaud, C. (2007) Rapport sur l'utilisation du néologisme « *bientraitance* » à propos de la protection des animaux. Available at: http://academieveterinaire.free.fr /rapports/bientraitance.pdf (accessed 23 June 2022).

Mills, A.D., Jones, R.B. and Faure, J.M. (1995) Species specificity of social reinstatement in Japanese quail *Coturnix japonica* genetically selected for high or low levels of social reinstatement behaviour. *Behavioural Processes* 34(1), 13–22. DOI: 10.1016/0376-6357(94)00044-h.

Molony, V. and Kent, J.E. (1997) Assessment of acute pain in farm animals using behavioral and physiological measurements. *Journal of Animal Science* 75(1), 266–272. DOI: 10.2527/1997.751266x.

Mormede, P., Boisseau-sowinski, L., Chiron, J., Diederich, C., Eddison, J. *et al.* (2018) Bien-être animal : contexte, définition, évaluation. *INRA Productions Animales* 31(2), 145–162. DOI: 10.20870/productions-animales.2018.31.2.2299.

Porcher, J. (2014) *Vivre Avec Les Animaux Une Utopie Pour Le XXIe Siècle.* La Découverte, 174 pp.

Sørensen, J.T. and Schrader, L. (2019) Labelling as a tool for improving animal welfare—the pig case. *Agriculture* 9(6), 123. DOI: 10.3390/agriculture9060123.

Thiermann, A. and Babcock, S. (2005) Animal welfare and international trade. *Revue Scientifique et Technique (International Office of Epizootics)* 24, 747–755.

Villain, A.S., Hazard, A., Danglot, M., Guérin, C., Boissy, A. *et al.* (2020) Piglets vocally express the anticipation of pseudo-social contexts in their grunts. *Scientific Reports* 10(1), 18496. DOI: 10.1038/s41598-020-75378-x.

Wiepkema, P.R., Van Hellemond, K.K., Roessingh, P. and Romberg, H. (1987) Behaviour and abomasal damage in individual veal calves. *Applied Animal Behaviour Science* 18(3–4), 257–268. DOI: 10.1016/0168-1591(87)90221-8.

Yeates, J.W. and Main, D.C.J. (2008) Assessment of positive welfare: a review. *Veterinary Journal* 175(3), 293–300. DOI: 10.1016/j.tvjl.2007.05.009.

General Conclusion

7

We have outlined many of the conceptual and practical advances in ethology. The serious ethologist approaches the analysis of animal behaviour with fascination and respect. Fascination and respect are what drives the acquisition of knowledge. 'Knowledge without compassion is inhumane, compassion without knowledge is ineffective' (Weisskopf, 1994, p. 128). This maxim is still relevant in 2022. Ethology, like all branches of science, is rigorously based on the formulating of hypotheses, the designing of experiments or observations to test these hypotheses and analysing the results of these tests to draw conclusions that can be scrutinized, challenged, and validated by peers. We emphasize particularly the rigorous challenging of hypotheses and conclusions. Approaching the behaviour of species scientifically allows one to develop an open, critical and undogmatic mind, and to be prepared to say, 'I don't know' or 'I was wrong'. We agree wholeheartedly with Tattersall (2002) who regretted that doubt was not taught sufficiently well in universities.

Ethologists have developed conceptual and methodological tools that help to understand how animals adapt themselves to make a living or how animals answer to the questions ethologists asked them in experiments. We consider that it is only with these tools, and with all the humility that the search for scientific knowledge imposes, that it is conceivable to speak for the animals. This involves patiently observing animals in structured conditions where the responses to the question are in a form that can be interpreted. Too often, people purport to speak on behalf of animals without ever having made the time or effort to observe them using such a scientific approach. They also proceed on unclear ground (Deputte and Vauclair, 1998).

We still have a lot to learn in view of the diversity of life around us and this should encourage us to learn more. It certainly should not be an excuse to postpone research in the field of animal welfare by blindly justifying production methods that may be threatened by the results. On the other side of the coin, there are many examples where the concepts

© CAB International 2022. *Interacting with Animals: Understanding their Behaviour and Welfare* (Pierre Le Neindre and Bertrand L. Deputte) DOI: 10.1079/9781800622418.0007

and results of ethology have been mobilized to enhance the productivity of animals, or to clarify issues in regulatory, legal and commercial fields. This puts a special responsibility on the shoulders of ethologists. Their results have important consequences not only for livestock farmers but also for all their fellow citizens. However, ethologists must distance themselves from the problems of commerce even if it is urgent to resolve them. This has been neatly summarized by Weisskopf (1994, p. 128): 'The basic sciences establish a link between humans and nature; they do not recognize industrial, national, racial and ideological limits. We, as a species, can only survive if science is used intensely and intelligently to improve the human condition.'

It is certainly important to be able to provide scientific answers to concrete, economic, zootechnical or other questions. We believe that we have provided enough evidence in this book to demonstrate that research with this as a goal must be based on a solid body of fundamental knowledge to encompass all the relationships in the life of any species. However, when relationships involve behaviour, we still lack this knowledge in many cases, particularly when it comes to farm animals. This is because animals are the product of natural selection. Over the course of evolution, they have acquired skills that have enabled them to adapt to the challenges they faced. For domestic species, this selection has been intensified by artificial selection and they have to confront additional challenges inherent in the constraints that humans impose on them. There are many such challenges and many ways to respond to them; and there is no single path to a satisfactory solution, but some paths are more effective than others. As behaviour is the combined result of phylogeny, which is set up during evolution, and ontogeny, which relates to the way an individual develops, only by exploring these two dimensions is it possible to obtain a sound analysis.

Over the period of an animal's life the expression of a diversity of behaviours will be required in response to both internal stimuli (such as hunger and mating) and external stimuli (like sociality, care for young, anti-predatory strategies and exploration). Behaviour is controlled by mechanisms ranging from the simplest reflexes and phylogenetically programmed responses, to the most complex that depend on learning and memory, culminating in the animal having consciousness and understanding. For most species, responses to new challenges result from their using cognitive skills that involve high plasticity of behaviour. We have highlighted recent scientific evidence that demonstrates the property of brain functioning in animals, such as consciousness, that were hitherto unrecognized. So we can now distinguish concepts like metacognition, self-awareness, episodic memory, future planning and theory of mind, all of which were previously thought of as being restricted to humans. But we only know about them in some species in some circumstances. We don't know whether these are the only species that possess these skills, or whether we have not yet succeeded in discovering them in other species and in other contexts. However, it is certainly no longer justifiable to claim that they are skills exclusive to humans or that they may not be complex in animals. Among gregarious species, we have highlighted the importance of group cohesion, which explains why individuals stay together and yet these animals may also be aggressive which seems to be contrary to sociality. However, aggressive behaviour in gregarious species is only regulatory to adjust the interactions of animals between themselves.

The livestock environment is an intricate ecosystem that fits into a complete chain of life in which humans are only one link. Recognizing and understanding the connections between all the links are important to conceptualize a coherent whole that integrates all aspects of life, from birth to death. Conceiving the world without the interactions between humans and animals, as vegans propose, seems to us a morbid mistake. Humans can turn to other humans to change their affiliation with nature, but nature remains deaf, and all ecosystems become fragile as soon as they are anthropized. To modify one component before having understood and analysed its role in this balance is like grasping a box from the bottom of a large pile and pretending that the pile won't collapse.

The co-adaptation of humans and animals on livestock farms has created both opportunities and constraints for animals. Moreover, both humans and animals have had to co-adapt in this environment. Protected by humans, the animal suffers less from unfavourable microclimates, parasites, poor food and disease. But, by contrast, it is confined, has reduced use of its environment, and cannot choose its diet and social partners. These constraints are the essence of the problem of maintaining animal welfare. We can only analyse this by studying how the animal chooses to cope with the situation. It makes its choices by exploiting its cognitive and emotional capacity – the same process that humans demonstrate. Just because a species has been domesticated does not automatically mean that humans completely control its behaviour. If selection by humans has focused on a behavioural trait, it will obviously be possible to control this trait better, but we should consider that a large part of free development depends on ontogenesis, or its development through its lifetime, and secondary phenotypic effects. So we are compelled to use all our creativity and scientific rigour to understand how animals express their behaviour in the face of the demands to which humans subject them, especially when these demands may seem selfish or unreasonable.

For each individual, each species and each environment, we find reactions that are anthropized. Even in non-agricultural environments, humans impose constraints that we cannot ignore. However, even if we would wish to do so, we can't give animals back their freedom because that could have disastrous consequences for them.

Control of reproduction and selection in animals often leads to profound genetic modification. Apart from taming them, we have imposed significant genetic modifications that have adapted them to the environments we impose but, more than that, we have also made them dependent on these environments. We have created animals that could not live without the support of humans and their methods, so these kind of animals could no longer reproduce or withstand elements of the natural environment, like extreme temperatures, parasites and predators. However, these animals are sometimes reduced to being simply objects to be valued according to our economic objectives. It is a trend that is surely detrimental to both humans and animals in the long term. One condition for a satisfactory life for animals is that they can choose, at least in part, the life they lead.

We will never know everything, but we now know enough to call for changes to current farming and supply-chain systems that genuinely have regard for the expectations of animals.

The economic optimization of livestock farms leads to a permanent increase in the constraints imposed on breeders and animals. Today's live-stock systems are largely the result of these attempts to optimize the use of resources, without providing for the quality of life of animals and those who breed them. Therefore, they sometimes interpose severe stresses into the living conditions of animals. Once, it was thought that suppressing violence towards and among animals was a desirable and sufficient goal to counter this. But this has ignored capitalizing on important positive behaviours, including those relating to the social life of animals, which we have high-lighted in this book. It also seems illusory to think that technical progress is the only way to solve what seems to be the squaring of the circle – in other words, ensuring decent living conditions for humans and animals, while keeping possibilities open for future improvements. With this in mind, we believe it is possible, and indeed necessary to revisit farming and supply-chain systems, and all aspects of animal life to ensure that we provide them with conditions in which to live satisfactorily. That is far from the case at present. Animals should always be considered as sentient beings and not as malleable objects. Systems that lead to mechanization and robotization, which are often now advocated, without doubt deleteriously loosen the links between humans and animals.

The main message of this book is that we have to listen to what animals have to say if we are to understand what is important to them. They are the ones who have the answers to the questions we ask them and we, in turn, must ask the right questions. We must not confine ourselves to thinking about what they can produce for us. We must question them relentlessly, even if it is difficult. We should not give in to the temptation to conclude that, since the actions of animals are too difficult to analyse and understand, we humans must simply decide among ourselves what we want to do for them. To repeat the words of Tinbergen (1953, p. 130): 'I must emphasize that thorough reading, while necessary, will never replace first-hand knowledge based on personal observations. Animals themselves are always more important than the books that have been written about them.'

We hope that this book will encourage readers to observe the animals we live with more closely and to question what they do and how we can persuade them to give us answers that are relevant. This is the only way to give us a sound base to devise new and more effective ways of living with them.

References

Deputte, B.L. and Vauclair, J. (1998) Connaître le monde, connaître les siens: les grands singes [n'] ont [pas] la parole. In: Bobbé, S. (ed.) *Grands Singes: la Fascination du Double. Collection Monde/Nature Extrême.* Editions Autrement, Paris, pp. 17–76.

Tattersall, I. (2002) *The Monkey in the Mirror.* Oxford University Press, Oxford, 220 pp.

Tinbergen, N. (1953) *Social Behaviour in Animals.* Methuen and Co, London, 188 pp.

Weisskopf, V.F. (1994) Essay: endangered support of basic science. *Scientific American* 270(5), 128. DOI: 10.1038/scientificamerican0594-128.

Index

Note: Page numbers in **bold** type refer to **figures**